Statistical Methods for Non-Precise DATA

Reinhard Viertl

Institute for Statistics and Probability Theory
Vienna University of Technology
(Technische Universität Wien)

CRC Press
Boca Raton   New York   London   Tokyo

Library of Congress Cataloging-in-Publication Data

Viertl, R. (Reinhard)
    Statistical methods for non-precise data / Reinhard Viertl.
      p.   cm.
    Includes bibliographical references and index.
    ISBN 0-8493-8242-4 (alk. paper)
    1. Fuzzy sets. 2. Mathematical statistics. I. Title.
QA248.V485   1995
    519.5′4--dc20   95-38137
                                                      CIP

No claim to original U.S. Government works
International Standard Book Number 0-8493-8242-4
Library of Congress Card Number 95-38137
Printed in the United States of America 1 2 3 4 5 6 7 8 9 0
Printed on acid-free paper

Author

Reinhard Viertl was born March 25, 1946, at Hall in Tyrol, Austria. He pursued studies in civil engineering and engineering mathematics, receiving his degree in engineering mathematics in 1972. He presented his doctoral thesis in mathematics and received his doctorate in engineering in 1974. Soon after, he was appointed assistant at the University of Technology Vienna and was promoted to University Docent in 1979. He was a research fellow and visiting lecturer at the University of California, Berkeley, from 1980 to 1981, and visiting lecturer at the University of Klagenfurt, Austria. Since 1982 Dr. Viertl has been full professor of applied statistics at the University of Technology Vienna. He is a Fellow of the Royal Statistical Society, London, held the Max Kade fellowship in 1980, and is founder of the Austrian Bayesian Society.

His book, *Statistical Methods in Accelerated Life Testing* appeared in 1988, and *Introduction to Stochastics* was published in German in 1990. He is editor of *Probability and Bayesian Statistics,* a volume that appeared in 1987, the book *Contributions to Environmental Statistics* (in German, 1992), and co-editor of a book titled *Mathematical and Statistical Methods in Artificial Intelligence* (1995).

Dr. Viertl is also the author of over 50 scientific papers in mathematics and statistics, most of which are in-depth examinations of algebra, probability theory, accelerated life testing, regional statistics, and statistics with non-precise data.

Preface

The formal description of non-precise data before their statistical analysis is, except from error models and interval arithmetic, a relatively young topic. Fuzziness is described in the theory of fuzzy sets but only few papers on statistical inference for non-precise data exist.

Especially in applied statistics, for example in environmetrics where very small concentrations have to be measured, it is necessary to describe the imprecision of data. Otherwise the results of statistical analyses can be very unrealistic and misleading.

For many statistical methods there exist generalizations to the situation of non-precise data given as non-precise numbers or non-precise vectors. These generalized inference methods are explained in this text.

The monograph is written for readers who are familiar with elementary statistical methods and simple stochastic models. The necessary knowledge is about an introduction to probability and statistics.

The goal of the text is to provide models for non-precise observations and methods of statistical inference for such data.

R. Viertl

Contents

Chapter I

Non-precise data and their formal description

1 Non-precise data

The results of measurements are often not precise real numbers or vectors but more or less *non-precise* numbers or vectors. This uncertainty is different from measurement errors and stochastic uncertainty and is called *imprecision*. Imprecision is a feature of single observations. Errors are described by statistical models and should not be confused with imprecision. In general imprecision and errors are superimposed. *In this monograph errors are not considered.*

Example 1.1: Many measurements in environmetrics are connected with a remarkable amount of uncertainty. For example the concentration of certain poisons in the air are non-precise quantities and their measurements are non-precise.

Example 1.2: The circumference of a tree is not a precise number. It is non-precise. In order to estimate the volume of a tree it is important to take care of the non-precise diameter.

Example 1.3: The lifetime of a system can in general not be described by a real number because the time of the end of the lifetime – for example a tree – is not precise.

Example 1.4: X-ray pictures are typical examples of non-precise data. These pictures are usually gray-tone pictures with more or less non-precise figures.

Example 1.5: Data from remote sensing are given as pictures from satellite observations. These are colored pictures which are very important in environmental research. Naturally these observations are non-precise to a certain degree.

Therefore the concept of *non-precise numbers* is necessary.

A special case of non-precise data is *interval data*.

Precise real numbers $x_0 \in \mathbb{R}$ as well as intervals $[a,b] \subseteq \mathbb{R}$ are uniquely characterized by their

indicator functions $I_{\{x_0\}}(\cdot)$ and $I_{[a,b]}(\cdot)$

respectively.

The indicator function $I_A(\cdot)$ of a classical set A is defined by

$$I_A(x) = \begin{cases} 1 & \text{for } x \in A \\ 0 & \text{for } x \notin A. \end{cases} \qquad (1.1)$$

The special indicator functions $I_{\{x_0\}}(\cdot)$ and $I_{[a,b]}(\cdot)$ from above are depicted in figure 1.1.

The imprecision of measurements implies that exact boundaries of interval data are not realistic. Therefore it is necessary to generalize real numbers and intervals to describe the imprecision.

This is done by the concept of *non-precise numbers* and general *non-precise subsets* of $\mathbb{R}$ as generalizations of intervals. Such non-precise subsets are also called *fuzzy sets*, which were introduced by L.A. Zadeh in 1963.

Non-precise numbers as well as non-precise subsets of $\mathbb{R}$ are described by generalizations of indicator functions, so-called *characterizing functions*.

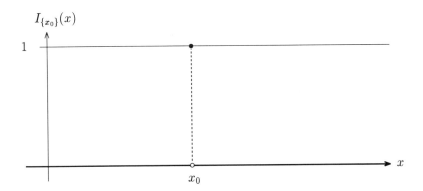

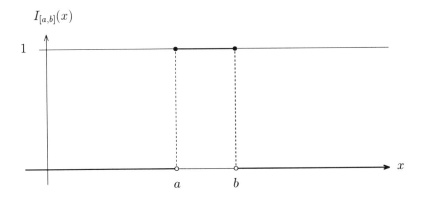

Figure 1.1: *Indicator functions*

Exercises

1. Let M be a fixed classical set. There is a one-to-one correspondence between classical sets A and indicator functions $I_A(\cdot)$. Moreover the following logical equivalence is valid:

$$A \subseteq B \Longleftrightarrow I_A(x) \leq I_B(x) \qquad \forall \ x \in M.$$

2. Show the following equations for subsets A and B of a fixed set M:

$$
\begin{aligned}
I_{A \cap B}(x) &= I_A(x) \cdot I_B(x) & \forall \ x \in M \\
I_{A \cup B}(x) &= I_A(x) + I_B(x) - I_A(x) \cdot I_B(x) & \forall \ x \in M \\
I_{A^c}(x) &= 1 - I_A(x) & \forall \ x \in M.
\end{aligned}
$$

3. Characterize the following non-precise lifetime observation. In engineering statistics, failures of equipment are recorded in the following way. Only the number of failures in certain time intervals $[t_i, t_i + \Delta t)$ are recorded. How can a non-precise observed lifetime of this type be described?

2 Non-precise numbers and characterizing functions

The modelling ideas for non-precise observations given in section 1 by so-called *non-precise numbers* $x^\star$ are quantified by so-called *characterizing functions* $\xi(\cdot)$, which characterize the imprecision of an observation. Such characterizing functions are generalizations of indicator functions.

Definition 2.1: A *characterizing function* $\xi(\cdot)$ of a non-precise number is a real function of a real variable with the following properties:

(1) $\xi : \mathbb{R} \to [0, 1]$

(2) $\exists\, x_0 \in \mathbb{R} : \xi(x_0) = 1$

(3) $\forall\, \alpha \in (0, 1]$ the set $B_\alpha := \{x \in \mathbb{R} : \xi(x) \geq \alpha\} = [a_\alpha, b_\alpha]$ is a finite closed interval, called *α-cut* of $\xi(\cdot)$.

The set $supp(\xi(\cdot)) := \{x \in \mathbb{R} : \xi(x) > 0\}$ is called *support* of $\xi(\cdot)$.

The set of all non-precise numbers is denoted by $\mathcal{F}(\mathbb{R})$.

In figure 2.1 some examples of characterizing functions are depicted.

In figure 2.2 the construction of an α-cut of a characterizing function is explained.

Remark 2.1: Non-precise observations and non-precise numbers will be marked by stars, i.e., $x^\star$, to distinguish them from (precise) real numbers x. Every non-precise number $x^\star$ is characterized by the corresponding characterizing function $\xi_{x^\star}(\cdot)$.

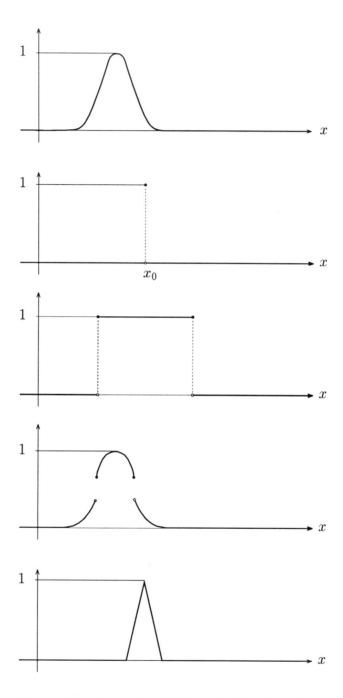

Figure 2.1: *Examples of characterizing functions*

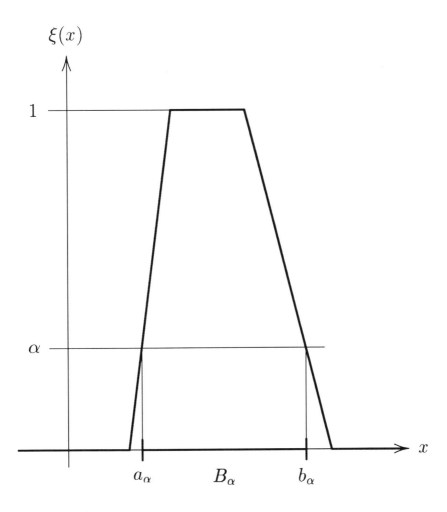

Figure 2.2: α-*Cut of a characterizing function*

Remark 2.2: Non-precise numbers are special fuzzy subsets of $\mathbb{R}$. This is reasonable as generalization of intervals, because intervals are classical subsets of $\mathbb{R}$.

Proposition 2.1: Characterizing functions $\xi(\cdot)$ are uniquely determined by the family

$$\big(B_\alpha;\ \alpha \in (0,1]\big)$$

of their α-cuts B_α and the following holds:

$$\xi(x) = \max_{\alpha \in (0,1]} \alpha \cdot I_{B_\alpha}(x) \qquad \forall\ x \in \mathbb{R}.$$

Proof : Let $x_0 \in \mathbb{R}$; then it follows

$$\alpha \cdot I_{B_\alpha}(x_0) = \alpha \cdot I_{\{x:\xi(x) \geq \alpha\}}(x_0) = \begin{cases} \alpha & \text{for} \quad \xi(x_0) \geq \alpha \\ 0 & \text{for} \quad \xi(x_0) < \alpha \end{cases}$$

and from that

$$\alpha \cdot I_{B_\alpha}(x_0) \leq \xi(x_0) \quad \forall\ \alpha \in (0,1]$$

and

$$\sup_{\alpha \in (0,1]} \alpha \cdot I_{B_\alpha}(x_0) \leq \xi(x_0).$$

For $\alpha_0 = \xi(x_0)$ we obtain

$$B_{\alpha_0} = \{x : \xi(x) \geq \xi(x_0)\} = [a_{\alpha_0}, b_{\alpha_0}]$$

and therefore

$$\alpha_0 \cdot I_{B_{\alpha_0}}(x_0) = \alpha_0 \cdot 1 = \xi(x_0) = \max_{\alpha \in (0,1]} \alpha \cdot I_{B_\alpha}(x_0).$$

$\square$

For characterizing functions of non-precise numbers fulfilling conditions (1) to (3) from definition 2.1, the following proposition holds.

Proposition 2.2: A real function $\xi : \mathbb{R} \to [0,1]$ which fulfills the following conditions:

(a) $\exists\ x_0 \in \mathbb{R} :\ \xi(x_0) = 1$

(b)　$\forall\ x_1 \in \mathbb{R}$ and $x_2 \in \mathbb{R}$ and $\lambda \in [0,1]$:
$$\xi\big(\lambda x_1 + (1-\lambda)x_2\big) \geq \min\big(\xi(x_1),\xi(x_2)\big)$$

(c)　$\forall\ x_0 \in \mathbb{R}$ and every sequence $x_n \to x_0$:
$$\lim_{x_n \to x_0} \xi(x_n) \leq \xi(x_0)$$

(d)　$\lim_{x \to -\infty} \xi(x) = 0$　and　$\lim_{x \to \infty} \xi(x) = 0$

is a characterizing function in the sense of definition 2.1 .

Proof: Condition (2) is equivalent to (a). For the α-cuts

$$B_\alpha = B_\alpha\big(\xi(\cdot)\big) = \{x \in \mathbb{R} :\ \xi(x) \geq \alpha\}$$

by condition (b) we have:
　　From $x_1 \in B_\alpha$ and $x_2 \in B_\alpha$ and $0 < \lambda < 1$ follows

$$\xi\big(\lambda x_1 + (1-\lambda)x_2\big) \geq \min\big(\xi(x_1),\xi(x_2)\big) \geq \alpha$$

and therefore
$$\lambda x_1 + (1-\lambda)x_2 \in B_\alpha.$$

From this it follows that B_α is an interval.

　　To see that B_α is a closed interval we use condition (c). For a limit point x_0 of B_α with $x_n \to x_0$ $(n \to \infty)$ we have

$$\xi(x_n) \geq \alpha \quad \forall\ n \in \mathbb{N},$$

and
$$\lim_{x_n \to x_0} \xi(x_n) \geq \alpha.$$

From condition (c) we obtain
$$\lim_{x_n \to x_0} \leq \xi(x_0)$$

and therefore
$$\xi(x_0) \geq \alpha.$$

By condition (d) the α-cuts are finite intervals for all $\alpha > 0$. Therefore condition (3) is fulfilled.

□

2.1   Examples of non-precise numbers

Non-precise numbers x^* can be described by corresponding characterizing functions $\xi_{x^*}(\cdot)$ in the following form

$$\xi_{x^*}(x) = \begin{cases} L(x) & \text{for} & x \leq m_1 \\ 1 & \text{for} & m_1 \leq x \leq m_2, \quad \text{with } m_1 \leq m_2 \\ R(x) & \text{for} & x \geq m_2 \end{cases}$$

where $L(\cdot)$ is a monotone increasing function and $R(\cdot)$ a decreasing real function.

The following special forms of characterizing functions are popular:

Trapezoid non-precise numbers

$$t^*(m_1, m_2, a_1, a_2)$$

$$L(x) \quad = \quad \max\left(0, \frac{x - m_1 + a_1}{a_1}\right)$$

$$R(x) \quad = \quad \max\left(0, \frac{m_2 + a_2 - x}{a_2}\right)$$

In figure 2.3 characterizing functions of trapezoid non-precise numbers are depicted.

Remark 2.3: For $m_1 = m_2$ we obtain so-called

triangular non-precise numbers

$$t^*(m, a_1, a_2).$$

Examples are given in figure 2.4.

Exponential non-precise numbers

$$e^*(m_1, m_2, a_1, a_2, p_1, p_2)$$

$$L(x) \quad = \quad \exp\left(-|\frac{x - m_1}{a_1}|^{p_1}\right)$$

$$R(x) \quad = \quad \exp\left(-|\frac{x - m_2}{a_2}|^{p_2}\right)$$

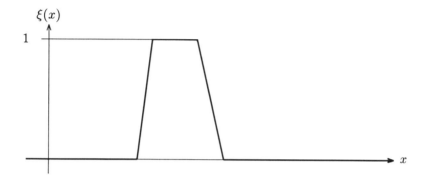

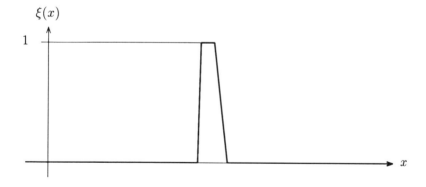

Figure 2.3: *Trapezoid non-precise numbers*

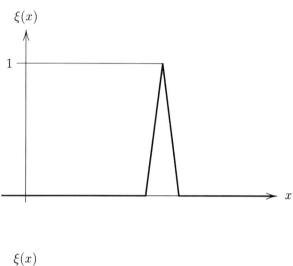

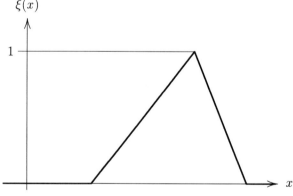

Figure 2.4: *Triangular non-precise numbers*

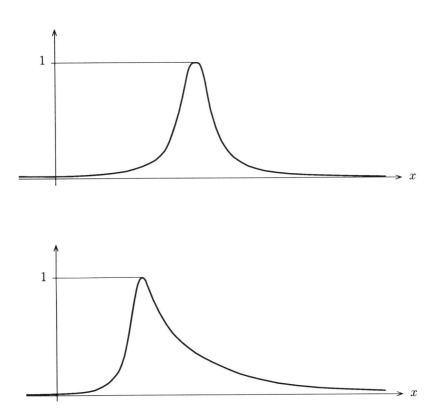

Figure 2.5: *Exponential non-precise numbers*

In figure 2.5 there are two examples of exponential non-precise numbers.

Application: Formal description of the non-precise quantity *water level* by a non-precise number. Here a non-precise number as depicted in figure 2.6 is suitable.

The problem of how to obtain the characterizing function of the quantity water level is discussed in section 3.

Example 2.1: The measurement of the concentration of a poison in the air at a fixed location and fixed time can be described by a trapezoid non-precise number.

2.2 Translation and scalar multiplication of non-precise numbers

Let $x^\star$ be a non-precise number with characterizing function $\xi(\cdot)$ and $c \in \mathbb{R}$. Then the non-precise number $x^\star + c$ is given by its characterizing function

$$\eta(x) = \xi(x - c) \quad \text{for all} \ \ x \in \mathbb{R}.$$

In a similar way the scalar multiplication of non-precise numbers is defined. For $c \in \mathbb{R}$ and $x^\star \,\hat{=}\, \xi(\cdot)$ the characterizing function $\psi(\cdot)$ of the non-precise number $c \cdot x^\star$ is given in the following way. For $c \neq 0$ it is defined by

$$\psi(x) = \xi\left(\frac{x}{c}\right) \quad \text{for all} \ \ x \in \mathbb{R}.$$

For $c = 0$ the characterizing function of $c \cdot x^\star$ is the indicator function $I_{\{0\}}(\cdot)$.

In figure 2.7 an example of a translated non-precise number is given and in figure 2.8 an example of a scalar multiplied non-precise number.

2.3 Convex hull of a non-convex fuzzy number

For functions $\varphi(\cdot)$ which fulfill conditions (1) and (2) of definition 2.1 but not condition (3) in the following way $\varphi(\cdot)$ can be transformed into a characterizing function $\xi(\cdot)$ fulfilling also condition (3).

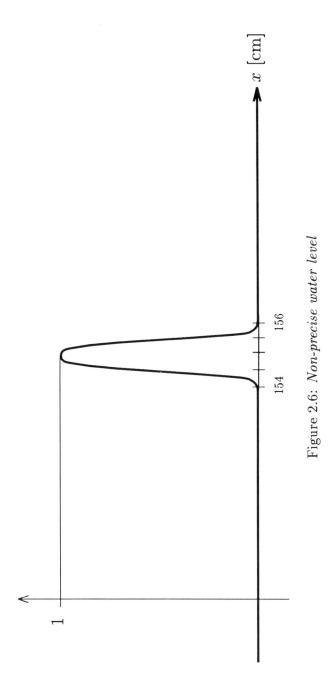

Figure 2.6: *Non-precise water level*

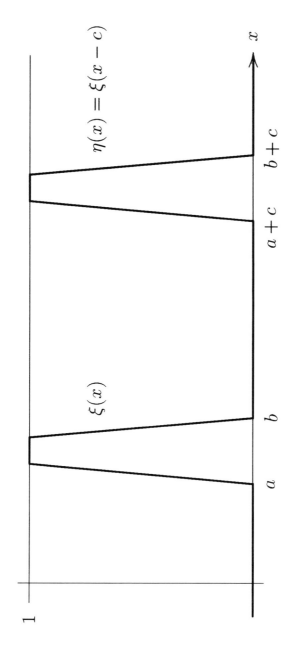

Figure 2.7: Translation of a non-precise number

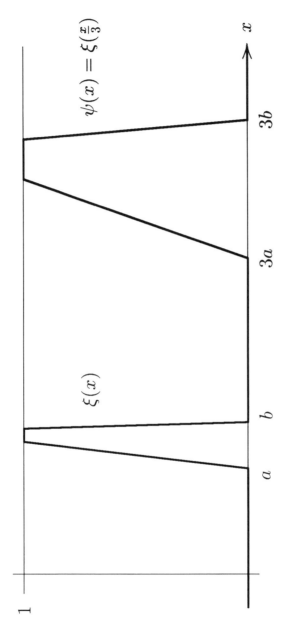

Figure 2.8: *Scalar multiplication of a non-precise number*

Definition 2.2: Let $\varphi(\cdot)$ be a real function fulfilling conditions (1) and (2) of definition 2.1 but not condition (3). Then some α-cuts B_α of $\varphi(\cdot)$ are unions of disjoint intervals $B_{\alpha,i}$, i.e., $B_\alpha = \bigcup\limits_{i=1}^{k_\alpha} B_{\alpha,i}$ with $B_{\alpha,i} = [a_{\alpha,i}; b_{\alpha,i}]$. Using proposition 2.1 the so-called *convex hull* $\xi(\cdot)$ of $\varphi(\cdot)$ is defined via the α-cuts $C_\alpha\big(\xi(\cdot)\big)$ by

$$C_\alpha := \left[\min_{i=1(1)k_\alpha} a_{\alpha,i};\ \max_{i=1(1)k_\alpha} b_{\alpha,i}\right].$$

Therefore $\xi(\cdot)$ is given by

$$\xi(x) = \max_{\alpha\in(0,1]} \alpha\cdot I_{C_\alpha}(x) \qquad \forall \quad x \in \mathbb{R}.$$

An example of constructing the convex hull is depicted in figure 2.9.

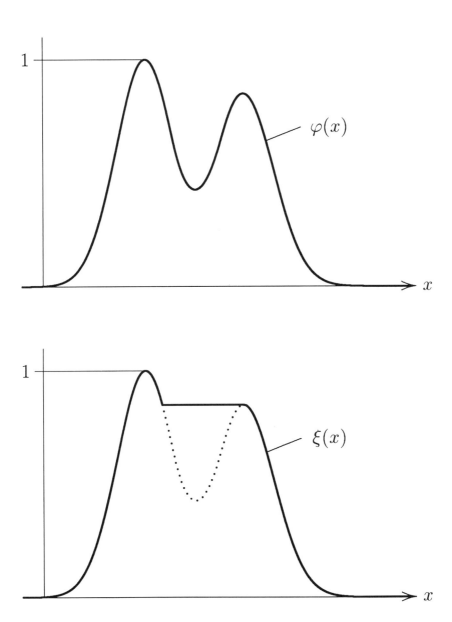

Figure 2.9: *Non-convex function $\varphi(\cdot)$ and its convex hull $\xi(\cdot)$*

Exercises

1. Let M be a classical set and $A \subseteq M$, and $I_A(\cdot)$ the indicator function of A. Determine the α-cut B_α of $I_A(\cdot)$ for all $\alpha \in [0, 1]$.

2. Determine the α-cuts of a characterizing function

$$\xi(x) = \begin{cases} L(x) & \text{for} \quad x < m_1 \\ 1 & \text{for} \quad m_1 \leq x \leq m_2 \\ R(x) & \text{for} \quad x > m_2 \end{cases}$$

where $L(\cdot)$ is a real and monotone increasing function and $R(\cdot)$ a real decreasing function with $\lim\limits_{x \downarrow -\infty} L(x) = 0$ and $\lim\limits_{x \uparrow \infty} R(x) = 0$.

3. Determine the α-cuts of an exponential non-precise number.

4. Take a characterizing function $\xi(\cdot)$ of a non-precise number $x^\star$ and calculate and draw the characterizing function of the non-precise number $x^\star + c$ for $c \in \mathbb{R}$. Also do the same for $c \cdot x^\star$. What is the relationship beween the α-cuts of $x^\star$ and the α-cuts of $x^\star + c$?

3 Construction of characterizing functions

The construction of characterizing functions to observed non-precise data depends on the field of application. Looking to real examples one can see some methodology.

Example 3.1: For the non-precise quantity water level of a river one can look at the intensity of wetness of the survey rod. Depicting the wetness as a function of the height as in figure 3.1 one can obtain the characterizing function of the quantity water level in the following way:

The dashed line in figure 3.1 is the derivative of the function $w(\cdot)$ which shows the wetness. Normalizing the dashed function by dividing it by its maximal value we obtain a characterizing function which describes the non-precise observation water level.

Example 3.2: Biological lifetimes are important examples of non-precise data. The observation of the lifetime of a tree is connected with this kind of uncertainty. Often the end of a lifetime of a system is characterized by the degradation of a certain quantity which is measured continuously. In figure 3.2 the method of obtaining the characterizing function $\xi(\cdot)$ of the non-precise lifetime is explained.

Remark 3.1: For functions $\xi(\cdot)$ obtained as in example 3.1 or similar results which have the shape depicted in figure 2.9 the *convex hull* of $\xi(\cdot)$ can be used. This convex hull $\eta(\cdot)$ is explained in section 2.3.

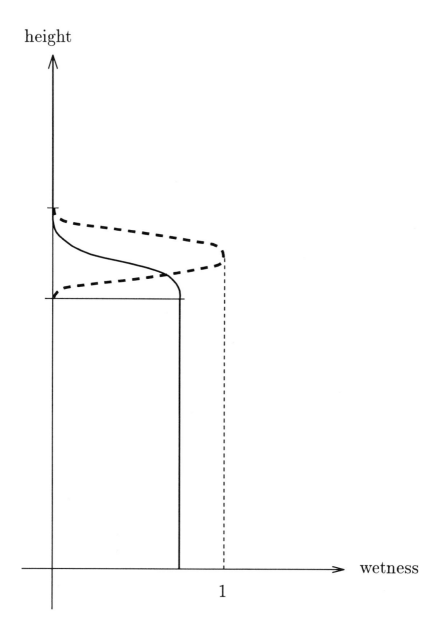

Figure 3.1: *Wetness of survey rod*

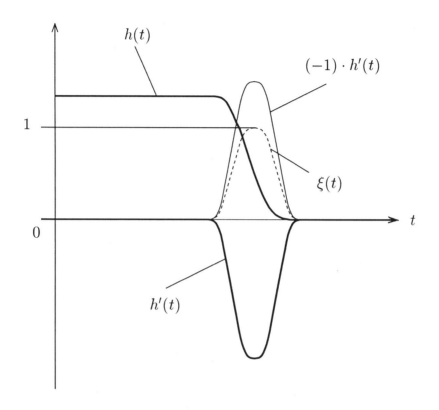

$$t \; \ldots \; \text{time}$$
$$h(t) \; \ldots \; \text{measured quantity}$$
$$h'(t) \; \ldots \; \text{derivative of } h(t)$$

$$\xi(t) = \frac{(-1)h'(t)}{\max|h'(t)|}$$

$\xi(\cdot)$ is the characterizing function

Figure 3.2: *Characterizing function of a non-precise lifetime*

Example 3.3: Many applied measurements are given as gray-tone pictures. An example is given in figure 3.3.

For one-dimensional quantities the intensity of the gray-tone is characteristic for the measurement. Depicting the intensity along a line gives an intensity function $h(\cdot)$ as in figure 3.4.

Remark 3.2: For intensity functions in form of step functions as in figure 3.5 the characterizing function can be obtained using the *rate of change*.

The rate of change $r(x)$ is given by a step function with constant values in the intervals

$$\left(x_i - \frac{\triangle x}{2}, \; x_i + \frac{\triangle x}{2} \right)$$

given by the intensity change at x_i.

$$r(x) = h(x_i + \delta) - h(x_i - \delta) \quad \text{for} \quad x \in \left(x_i - \frac{\triangle x}{2}, \; x_i + \frac{\triangle x}{2} \right)$$

The characterizing function $\xi(\cdot)$ is the *scaled rate of change*.

$$\xi(x) = \frac{r(x)}{\max_{x \in \mathbb{R}} r(x)}.$$

In case the function $\xi(\cdot)$ is a non-convex fuzzy number, the convex hull of $\xi(\cdot)$ is taken as the characterizing function.

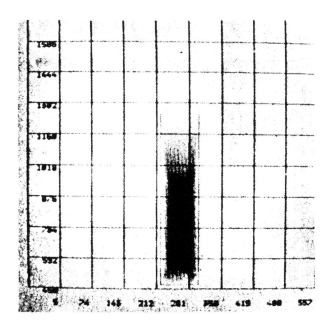

One observation as presented by the screen

Figure 3.3: *Gray-tone picture*

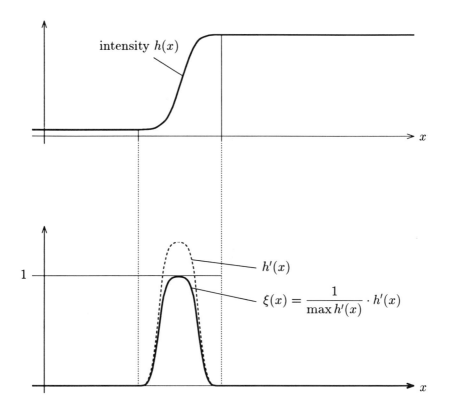

Figure 3.4: *Gray-intensity along a line and construction of the characterizing function*

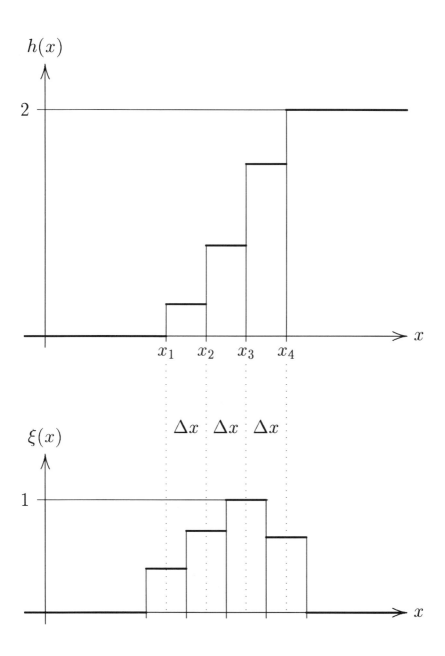

Figure 3.5: *Gray-intensity $h(x)$ in form of a step function and scaled rate of change*

Exercises

1. Draw the characterizing function for a gray-tone intensity given in the following picture. Assume a trapezoid characterizing function.

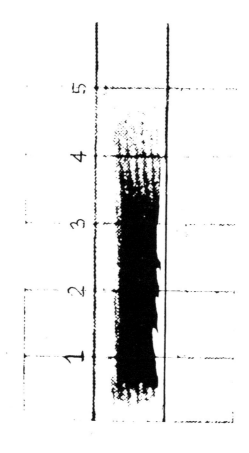

2. Let the gray-tone intensity of a measurement be given by
the following diagram

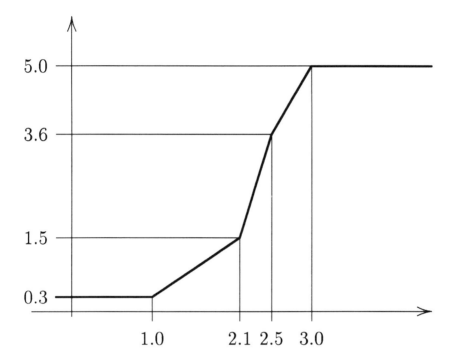

Construct the characterizing function of this non-precise value of a one-dimensional gray-intensity.

4   Non-precise vectors

Many statistical data are — in case of idealized precise observations — given by vectors of real numbers. Real multi-dimensional measurement data are mostly non-precise. Therefore this imprecision has to be described.

Example 4.1: Measuring the location of a point in the plane one obtains an idealized pair $(x, y) \in \mathbb{R}^2$. For real observations one obtains — for example by a radar device — a light point with finite range. This is a non-precise two-dimensional vector $\underline{x}^\star$ with *characterizing function* $\xi_{\underline{x}^\star}(x_1, x_2)$ which describes the imprecision of the observation. The values $\xi(x_1, x_2)$ are given by the intensity of the light point (Compare figure 4.1).

For the mathematical description of vector-valued non-precise observations one can use so-called *non-precise vectors* which are described by *characterizing functions* also.

Definition 4.1: A non-precise n-dimensional vector $\underline{x}^\star$ is determined by a corresponding *characterizing function* $\xi_{\underline{x}^\star}(\cdot)$ of n real variables with the following properties:

(1) $\xi_{\underline{x}^\star} : \mathbb{R}^n \to [0, 1]$

(2) $\exists\, \underline{x}_0 \in \mathbb{R}^n : \xi_{\underline{x}^\star}(\underline{x}_0) = 1$

(3) $B_\alpha(\underline{x}^\star) := \{\underline{x} \in \mathbb{R}^n : \xi_{\underline{x}^\star}(\underline{x}) \geq \alpha\}$ is $\forall\, \alpha \in (0, 1]$
     a simply connected and compact subset of $\mathbb{R}^n$.

The set $supp\left(\xi_{\underline{x}^\star}(\cdot)\right) := \{\underline{x} \in \mathbb{R}^n : \xi_{\underline{x}^\star}(\underline{x}) > 0\}$ is called *support* of $\xi_{\underline{x}^\star}(\cdot)$.

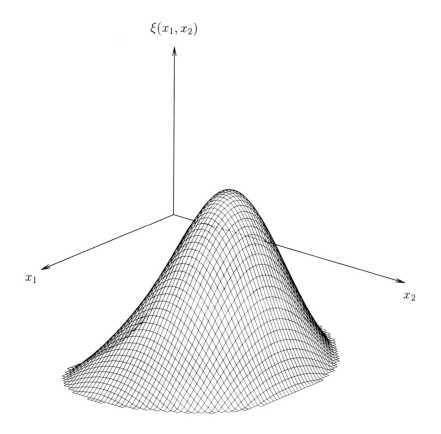

The values $\xi(x_1, x_2)$ are given by the intensity of the light "point"

Figure 4.1: *Non-precise vector*

The set of all non-precise n-dimensional vectors is denoted by $\mathcal{F}(\mathbb{R}^n)$.

Proposition 4.1: Characterizing functions $\xi_{\underline{x}^\star}(\cdot)$ of non-precise vectors $\underline{x}^\star$ are determined by their families

$$\left(B_\alpha(\underline{x}^\star); \quad \alpha \in (0,1]\right)$$

of α-cuts and the following is valid:

$$\xi_{\underline{x}^\star}(\underline{x}) = \max_{\alpha \in (0,1]} \alpha \cdot I_{B_\alpha(\underline{x}^\star)}(\underline{x}) \qquad \forall \quad \underline{x} \in \mathbb{R}^n.$$

Proof: For $\underline{x}_0 \in \mathbb{R}^n$ and $\forall \, \alpha \in (0,1]$ we have

$$\alpha \cdot I_{B_\alpha(\underline{x}^\star)}(\underline{x}_0) = \alpha \cdot I_{\{\underline{x} \,:\, \xi_{\underline{x}^\star}(\underline{x}) \geq \alpha\}}(\underline{x}_0) = \begin{cases} \alpha & \text{for} \quad \xi_{\underline{x}^\star}(\underline{x}_0) \geq \alpha \\ 0 & \text{for} \quad \xi_{\underline{x}^\star}(\underline{x}_0) < \alpha. \end{cases}$$

From that we obtain

$$\alpha \cdot I_{B_\alpha(\underline{x}^\star)}(\underline{x}_0) \leq \xi_{\underline{x}^\star}(\underline{x}_0) \quad \forall \, \alpha \in (0,1]$$

and therefore

$$\sup_{\alpha \in (0,1]} \alpha \cdot I_{B_\alpha(\underline{x}^\star)}(\underline{x}_0) \leq \xi_{\underline{x}^\star}(\underline{x}_0).$$

For $\alpha_0 = \xi_{\underline{x}^\star}(\underline{x}_0)$ it follows

$$B_{\alpha_0}(\underline{x}^\star) = \{\underline{x} \,:\, \xi_{\underline{x}^\star}(\underline{x}) \geq \xi_{\underline{x}^\star}(\underline{x}_0)\}$$

and

$$I_{B_{\alpha_0}(\underline{x}^\bullet)}(\underline{x}_0) = 1$$

and moreover

$$\alpha_0 \cdot I_{B_{\alpha_0}(\underline{x}^\bullet)}(\underline{x}_0) = \alpha_0 = \xi_{\underline{x}^\star}(\underline{x}_0) = \max_{\alpha \in (0,1]} \alpha \cdot I_{B_\alpha(\underline{x}^\star)}(\underline{x}_0).$$

$\square$

Proposition 4.2: A measurable function $\xi : \mathbb{R}^n \to [0,1]$ with the following properties

(a) $\quad \exists \, \underline{x}_0 \in \mathbb{R}^n : \quad \xi(\underline{x}_0) = 1$

(b) $\forall\ \underline{x}_1 \in \mathbb{R}^n$ and $\underline{x}_2 \in \mathbb{R}^n$ with $\lambda \in (0,1)$ it follows:
$$\xi\big(\lambda\underline{x}_1 + (1-\lambda)\underline{x}_2\big) \geq \min\big(\xi(\underline{x}_1), \xi(\underline{x}_2)\big)$$

(c) $\forall\ \underline{x}_0 \in \mathbb{R}^n$ and every sequence $\underline{x}_n \to \underline{x}_0$ it follows:
$$\lim_{\underline{x}_n \to \underline{x}_0} \xi(\underline{x}_n) \leq \xi(\underline{x}_0)$$

(d) $\displaystyle\lim_{|\underline{x}| \to \infty} \xi(\underline{x}) = 0$

is a characterizing function in the sense of definition 4.1.

Proof: Properties (1) and (2) from definition 4.1 are trivially fulfilled. Property (3) is seen in a similar way as in the proof of proposition 2.2.

□

Remark 4.1: The construction of the characterizing function of a non-precise vector in the example at the beginning of this section is the following. Take the light intensity $h(x_1, x_2)$ at every coordinate vector $(x_1, x_2) \in \mathbb{R}^2$ and normalize it, i.e.,

$$\xi(x_1, x_2) = \frac{h(x_1, x_2)}{\max\limits_{(x_1,x_2) \in \mathbb{R}^2} h(x_1, x_2)} \qquad \text{for all } (x_1, x_2) \in \mathbb{R}^2.$$

Exercises

1. Consider the situation where the result of one observation is given by a two-dimensional gray-tone picture with discrete gray-intensities. This is the generalization of figure 3.5 for two-dimensional data. How has the scaled rate of change to be generalized to obtain a non-precise vector?

2. Let $\xi_1(\cdot)$ and $\xi_2(\cdot)$ be two characterizing functions of non-precise numbers. Explain that the function $\xi(\cdot, \cdot)$ of two real variables defined by

$$\xi(x_1, x_2) := \min\big(\xi_1(x_1), \xi_2(x_2)\big) \qquad \forall\ (x_1, x_2) \in \mathbb{R}^2$$

is the characterizing function of a non-precise vector.

5 Functions of non-precise quantities and non-precise functions

In statistics often functions of observations are important. An example is the sample mean $\overline{x}_n$ of n observations $x_1, \ldots, x_n$ of a stochastic quantity.

In case of non-precise observations $x_1^\star, \ldots, x_n^\star$ we have a function of non-precise quantities and the value of the function is a non-precise element of the domain.

More generally non-precise functions can be described as generalizations of classical functions $g : M \to \mathbb{R}$ in the following way.

Definition 5.1: A *non-precise function* $g^\star(\cdot)$ is a mapping which assigns to every element $x \in M$ a non-precise number $y_x^\star = g^\star(x) \in \mathcal{F}(\mathbb{R})$.

Remark 5.1: Non-precise functions are uniquely given by a family

$$\left(\phi_x(\cdot); \; x \in M \right)$$

of non-precise numbers $y_x^\star$ with corresponding characterizing functions $\phi_x(\cdot)$. For graphical presentations of non-precise functions the description of so-called α-*level curves* is of advantage. These α-level curves are defined as follows.

For $\alpha \in (0, 1]$ consider the α-cuts

$$B_\alpha(y_x^\star) = \left[\underline{g}_\alpha(x), \; \overline{g}_\alpha(x) \right] \quad \text{for every} \;\; x \in M.$$

Then for variable x two classical functions $\underline{g}_\alpha(x)$ and $\overline{g}_\alpha(x)$ are obtained.

The graphs of these functions are called α-*level curves* of the non-precise function $g^*(\cdot)$. An example for $M = \mathbb{R}$ is given in figure 5.1.

Another kind of generation of imprecision of functions is the one induced by the imprecision of argument values of a function.

Definition 5.2: Let $g : \mathbb{R}^n \to \mathbb{R}$ be a classical function whose argument values $\underline{x}$ are not known precisely but are non-precise vectors $\underline{x}^\star \in \mathcal{F}(\mathbb{R}^n)$ with corresponding characterizing function $\xi(\cdot)$. Then a non-precise value $g(\underline{x}^\star)$ of the function with characterizing function $\psi(\cdot)$ is given by

$$\psi(y) := \sup\{\xi(\underline{x}) : \underline{x} \in \mathbb{R}^n \wedge g(\underline{x}) = y\}.$$

If no $\underline{x}$ exists with $g(\underline{x}) = y$ then $\psi(y) = 0$.

Remark 5.2: For general functions $g(\cdot)$ it is not provided that $\psi(\cdot)$ is a characterizing function in the sense of definition 2.1. For *continuous functions* the conditions are fulfilled. This is shown in the following proposition.

Proposition 5.1: Let $\underline{x}^\star \in \mathcal{F}(\mathbb{R}^n)$ be a non-precise vector with characterizing function $\xi(\cdot)$ and $g : M \to \mathbb{R}$ a continuous function with $M \subseteq \mathbb{R}^n$ and $supp(\xi(\cdot)) \subseteq M$. Then the function $\psi(\cdot)$ from definition 5.2 is a characterizing function in the sense of definition 2.1 which defines a non-precise number y^*. For the α-cut presentation

$$\left(B_\alpha(y^*); \ \alpha \in (0, 1]\right)$$

of $y^\star$ we obtain the following intervals

$$B_\alpha(y^*) = \left[\min_{\underline{x} \in B_\alpha(\underline{x}^\star)} g(\underline{x}), \ \max_{\underline{x} \in B_\alpha(\underline{x}^\star)} g(\underline{x}) \right].$$

Proof: The conditions (1) to (3) of characterizing functions from definition 2.1 have to be shown. The condition in (1) and condition (2) are trivially fulfilled. By the continuity of $g(\cdot)$ the set $g^{-1}(\{y\})$ is closed and therefore we obtain

$$\sup\{\xi(\underline{x}) : \underline{x} \in g^{-1}(\{y\})\} = \max\{\xi(\underline{x}) : \underline{x} \in g^{-1}(\{y\})\}.$$

Next we show

$$B_\alpha(y^*) = g\left(B_\alpha(\underline{x}^*)\right) \quad \forall \ \alpha \in (0, 1].$$

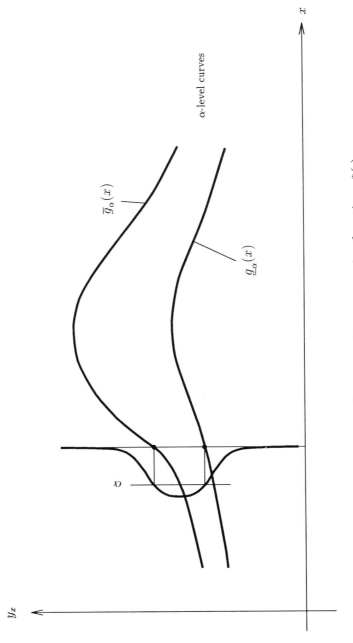

Figure 5.1: α-Level curves of a non-precise function $g^*(\cdot)$

For $\alpha \in (0,1]$ and $y \in g\big(B_\alpha(\underline{x}^\star)\big)$ there exists an $\underline{x} \in B_\alpha(\underline{x}^\star)$ with $y = g(\underline{x})$. From that follows $\xi(\underline{x}) \geq \alpha$ and

$$\sup\{\xi(\underline{x}) : g(\underline{x}) = y\} \geq \alpha,$$

and by definition 5.2 $\psi(y) \geq \alpha$ which yields $y \in B_\alpha(y^\star)$, and from that we obtain $g\big(B_\alpha(\underline{x}^\star)\big) \subseteq B_\alpha(y^\star)$. On the other side for $y \in B_\alpha(y^\star)$ we have $\psi(y) \geq \alpha$ and

$$\sup \{\xi(\underline{x}) : g(\underline{x}) = y\} \geq \alpha.$$

By the continuity of $g(\cdot)$ it follows

$$\max\{\xi(\underline{x}) : g(x) = y\} = \sup\{\xi(\underline{x}) : g(\underline{x}) = y\} \geq \alpha.$$

Therefore there exists an $\underline{x}_0 : \xi(\underline{x}_0) \geq \alpha$ with $g(\underline{x}_0) = y$ and $\underline{x}_0 \in B_\alpha(\underline{x}^\star)$. From this follows $y \in g\big(B_\alpha(\underline{x}^\star)\big)$ and $B_\alpha(y^\star) \subseteq g\big(B_\alpha(\underline{x}^\star)\big)$.

By the continuity of $g(\cdot)$ and the fact that $B_\alpha(\underline{x}^\star)$ is a compact and simply connected subset of $\mathbb{R}^n$ it follows that $g\big(B_\alpha(\underline{x}^\star)\big)$ is a compact and simply connected subset of $\mathbb{R}$ and therefore a closed interval of the form given in the proposition.

$$\square$$

Remark 5.3: In general it can be complicated to obtain the analytical form of $\psi(\cdot)$ from definition 5.2. For linear functions

$$g(\underline{x}) = A \, \underline{x} = \sum_{i=1}^{n} a_i x_i$$

with $A = (a_1, \ldots, a_n)$, $a_i \in \mathbb{R}$, this is simple (compare exercise 2 in section 5).

For general functions the imprecision of the result can be graphically represented by a finite number of α-cuts of $\psi(\cdot)$. An example is given in figure 5.2 where ten α-cuts are depicted.

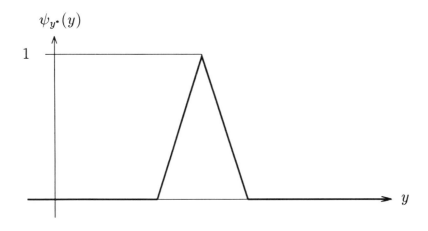

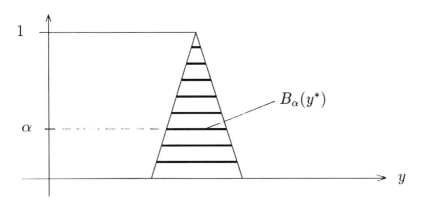

Figure 5.2: *Characterizing function and corresponding*
α-cuts

Exercises

1. Work out the difference between non-precise functions and functions of non-precise arguments.

2. Let $g : \mathbb{R}^n \to \mathbb{R}$ with vector $A = (a_1, \ldots, a_n)$ such that $g(\underline{x}) = A\underline{x}$ $\forall \underline{x} \in \mathbb{R}^n$. Prove that the function $\psi(\cdot)$ from definition 5.2 is a characterizing function for all non-precise vectors $\underline{x}^\star \in \mathcal{F}(\mathbb{R}^n)$.

3. Non-precise functions in the sense of definition 5.1 can be used as mathematical description of fuzzy *a priori* information in Bayesian statistical inference. Use this for fuzzy *a priori* information in sampling from a Bernoulli experiment. Note that the beta-distributions form a conjugate *a priori* family for the Bernoulli experiment.

Chapter II

Descriptive statistics with non-precise data

6 Non-precise samples

A sample of a stochastic quantity X is n observations $x_1, \cdots, x_n$ of X. The considered quantities can be one-dimensional, i.e., $x_i \in \mathbb{R}$ or of higher dimension, i.e., $x_i \in \mathbb{R}^k$.

6.1 One-dimensional quantities

Observations of a one-dimensional stochastic quantity (also called random quantity) are often non-precise. Such data are described by non-precise numbers x^* and n observations by n non-precise numbers

$$x_1^\star, \ldots, x_n^\star$$

with corresponding characterizing functions

$$\xi_1(\cdot), \ldots, \xi_n(\cdot).$$

These n observations are also called *non-precise sample* or *non-precise data* and are denoted by $D^\star$.

Remark 6.1: There can be also precise observations $x_i \in \mathbb{R}$ in the sample. In this case the corresponding characterizing function is the one point indicator function $I_{\{x_i\}}(\cdot)$.

For one-dimensional quantities non-precise samples are finite sequences of non-precise numbers. The characterizing functions of the non-precise numbers can have intersecting supports. An example is given in figure 6.1.

Some methods of descriptive statistics use functions of the sample. This is no problem in case of precise observations. In case of precise data $D = (x_1, \ldots, x_n)$ with observation space M_X we

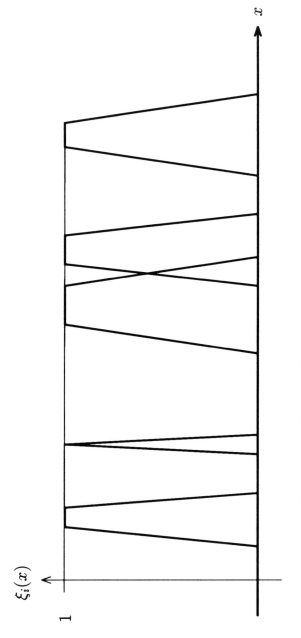

Figure 6.1: *Characterizing functions of a non-precise sample*

have $x_i \in M_X$. The observations are put together into a vector $\underline{x} = (x_1, \ldots, x_n)$ which is an element of the *sample space* M_X^n.

For non-precise observations it is necessary to generalize functions and definition 5.2 from section 5 can be used. Details are explained in section 11.

Also elementary concepts such as histograms, cumulative sums, and empirical distribution functions need generalizations.

6.2 Samples from stochastic vectors

In this case the observation space M_X is a subset of $\mathbb{R}^k$ and the sample space $M_X^n \subseteq (\mathbb{R}^k)^n = \mathbb{R}^{n \cdot k}$.

For precise data $\underline{x}_1, \ldots, \underline{x}_n$ the observations are put together and we obtain the so-called *data matrix*

$$
\begin{matrix}
x_{11} & x_{12} & \cdots & x_{1n} \\
x_{21} & x_{22} & \cdots & x_{2n} \\
\vdots & \vdots & \ddots & \vdots \\
x_{k1} & x_{k2} & \cdots & x_{kn}
\end{matrix}
$$

which can be considered as an element of $\mathbb{R}^{n \cdot k}$.

In case of non-precise vector-observations $\underline{x}_i^\star$ with corresponding characterizing functions $\xi_i(\cdot)$ of these non-precise vectors $\underline{x}_i^\star$ the observations have to be combined to form a non-precise element of the sample space. Details on that are explained in section 11.2.

Exercises

1. Let $x_1^\star$ and $x_2^\star$ be two non-precise observations of a one-dimensional stochastic quantity X with observation space M_X and the corresponding characterizing functions $\xi_1(\cdot)$ and $\xi_2(\cdot)$. Then a fuzzy element of the sample space M_X^2 is given by the function

$$
\varphi(x_1, x_2) := \xi_1(x_1)\xi_2(x_2) \qquad \forall \ (x_1, x_2) \in M_X^2.
$$

Explain the relationship to exercise 2 in section 4 and draw the results for two characterizing functions of your choice.

2. For two non-precise observations $\underline{x}_1^\star$ and $\underline{x}_2^\star$ of a two-dimensional stochastic vector $\underline{X} = (X, Y)$ with two-dimensional characterizing functions $\xi_1(\cdot)$ and $\xi_2(\cdot)$ the characterizing function $\varphi(\cdot)$ of a corresponding fuzzy element of the sample space $\mathbb{R}^4$ can be defined by

$$\xi(x_1, y_1, x_2, y_2) := \min\left(\xi_1(x_1, y_1),\ \xi_2(x_2, y_2)\right).$$

An alternative approach is

$$\varphi(x_1, y_1, x_2, y_2) := \xi_1(x_1, y_1)\xi_2(x_2, y_2)$$

$$\forall\ \ (x_1, y_1, x_2, y_2) \in \mathbb{R}^4.$$

What is the relationship between the two definitions? Show that for interval data the results are identical.

7 Histograms for non-precise data

The construction of a histogram based on non-precise data makes problems because, for a non-precise observation, it is in general not uniquely decidable if the observation belongs to a given class K_j. This ambiguity occurs if the support of the characterizing function of the observation has non-empty intersection with K_j and also non-empty intersection with the complement K_j^c. The situation is depicted in figure 7.1.

For this section all characterizing functions $\xi(\cdot)$ are assumed to fulfill

$$\int_{\mathbb{R}} \xi(x)dx < \infty.$$

If $supp\big(\xi(\cdot)\big)$ has non-empty intersection with two classes it is not possible to decide that the observation belongs to a certain class.

In general for n non-precise observations there exists no exact height $h_n(K_j)$ of the histogram. Therefore it is necessary to construct a generalization, so-called *fuzzy histograms*. In such generalized histograms the height over a class K_j is a non-precise element of the set

$$\left\{0, \frac{1}{n}, \frac{2}{n}, \ldots, \frac{n-1}{n}, 1\right\}.$$

Definition 7.1: For a non-empty half-ordered set $(M, \leq)$ a *non-precise element* in M is given by a function $\varphi : M \to [0,1]$ with the following properties:

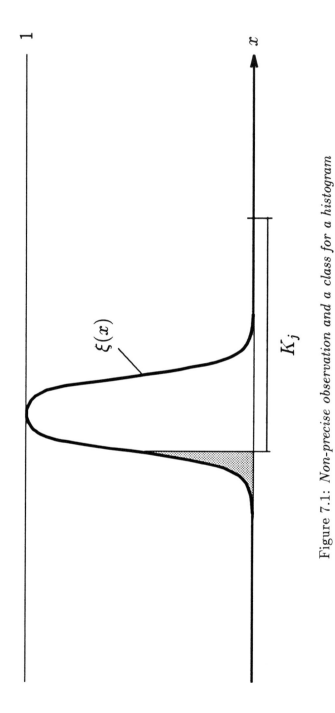

Figure 7.1: *Non-precise observation and a class for a histogram*

(1) $\quad\exists\, x \in M : \varphi(x) = 1$

(2) $\quad\varphi(x) \geq \min\{\varphi(u), \varphi(v)\}$ for $u \leq x \leq v$ and $\{x, u, v\} \subseteq M$

(3) $\quad$In case of continuous space M, for all $x \in M$ and
$$x_n \to x \;:\; \lim_{x_n \to x} \varphi(x_n) \leq \varphi(x)$$

The function $\varphi(\cdot)$ is called *characterizing function* of the non-precise element.

For n non-precise observations $x_1^\star, \ldots, x_n^\star$ with corresponding characterizing functions $\xi_1(\cdot), \ldots, \xi_n(\cdot)$ and every class K_j of a decomposition of the observation space a

$$\textit{non-precise\ \ relative\ \ frequency}\ \ h_n^\star(K_j)$$

is obtained in the following way: The non-precise relative frequencies are non-precise elements of the ordered set $\{0, \frac{1}{n}, \frac{2}{n}, \ldots, \frac{n-1}{n}, 1\}$, whose characterizing functions

$$\varphi_{h_n^\star(K_j)}(\cdot)$$

are obtained from the characterizing functions

$$\varphi_{n_{K_j}^\star}(\cdot) \text{ of the } \textit{non-precise absolute frequencies } n_{K_j}^\star.$$

For $n_{K_j}^\star$ we obtain as lower bound $\underline{n}_{K_j}$ the number of observations $x_i^\star$, for which $supp\big(\xi_i(\cdot)\big) \subseteq K_j$, i.e.,

$$\underline{n}_{K_j} \;=\; \text{number of } x_i \text{ with } supp\big(\xi_i(\cdot)\big) \subseteq K_j.$$

The upper bound $\overline{n}_{K_j}$ is the number of observations $x_i^\star$ for which $supp\big(\xi_i(\cdot)\big)$ has non-empty intersection with K_j, e.g.,

$$\overline{n}_{K_j} \;=\; \text{number of } x_i \text{ with } supp\big(\xi_i(\cdot)\big) \cap K_j \neq \emptyset.$$

The values $\varphi_{n_{K_j}^\star}(l)$ of the characterizing function $\varphi_{n_{K_j}^\star}(\cdot)$ can be obtained in the following way.

Let $x_{l_j}^\star$, $l_j \in \{1, \cdots, n\}$ be the non-precise observations in the sample $x_1^\star, \cdots, x_n^\star$ with

$$supp\big(\xi_i(\cdot)\big) \cap K_j \neq \emptyset \ \text{ but } \ supp\big(\xi_i(\cdot)\big) \not\subseteq K_j.$$

Let n_j be the number of these observations; then, we have

$$n_j = \overline{n}_{K_j} - \underline{n}_{K_j}.$$

In case the classes K_j are intervals, i.e., $K_j = [a_j, b_j]$ we define

$$A_j(x_{l_j}^*) := \int_{a_j}^{b_j} \xi_{l_j}(x)dx$$

and

$$B_j(x_{l_j}^*) := \int_{\mathbb{R}} \xi_{l_j}(x)dx - A_j(x_{l_j}^*). \tag{7.1}$$

These areas are used to obtain *non-precise absolute frequencies*

$$n_{K_j}^* \text{ of } K_j \text{ for all classes } K_j, \quad j = 1(1)k.$$

Ordering the numbers $B_j(x_{l_j}^*)$ in increasing order we obtain the finite sequence

$$B_j^{(i)}(x_{l_j}^*), \quad i = 1(1)n_j.$$

$A_j^{(i)}(x_{l_j}^*)$ is the corresponding area from equation (7.1).

These values are used to construct the characterizing function $\varphi_{n_{K_j}^*}(\cdot)$ of the non-precise absolute frequency $n_{K_j}^*$ of the class K_j. The values of this characterizing function are given by

$$\varphi_{n_{K_j}^*}(m) = 0 \qquad \forall \; m < \underline{n}_{K_j}$$
$$\varphi_{n_{K_j}^*}(\underline{n}_{K_j}) = 1$$
$$\vdots$$
$$\varphi_{n_{K_j}^*}(\underline{n}_{K_j} + s) = 1 - \frac{1}{n}\sum_{i=1}^{s} \frac{B_j^{(i)}(x_{l_j}^*)}{B_j^{(i)}(x_{l_j}^*) + A_j^{(i)}(x_{l_j}^*)}$$
$$\vdots$$
$$\varphi_{n_{K_j}^*}(\underline{n}_{K_j} + n_j) = \varphi_{n_{K_j}^*}(\overline{n}_{K_j}) = 1 - \frac{1}{n}\sum_{i=1}^{n_j} \frac{B_j^{(i)}(x_{l_j}^*)}{B_j^{(i)}(x_{l_j}^*) + A_j^{(i)}(x_{l_j}^*)}$$
$$\varphi_{n_{K_j}^*}(m) = 0 \qquad \forall \; m > \overline{n}_{K_j}.$$

The characterizing function $\eta_j(\cdot)$ of the non-precise height $h_n^\star(K_j)$ over the class K_j is given by

$$\eta_j\left(\frac{l}{n}\right) = \varphi_{n_{K_j}^\star}(l) \quad \text{for } l \in \{0, 1, 2, \ldots, n\}.$$

For every class K_j the lower bound of the non-precise height is

$$\underline{h}_n(K_j) = \frac{\underline{n}_{K_j}}{n}$$

and the upper bound is

$$\overline{h}_n(K_j) = \frac{\overline{n}_{K_j}}{n}.$$

In figure 7.2 the non-precise value $h_n^\star(K_j)$ and an element of the fuzzy histogram is depicted.

Remark 7.1: There are also other generalizations of histograms possible in connection with the so-called extension principle from fuzzy set theory. For details compare [17] from the references.

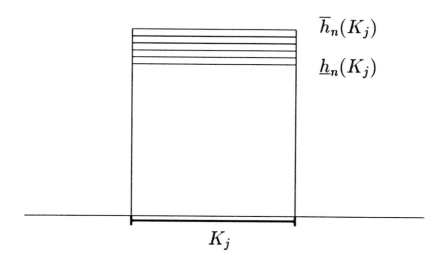

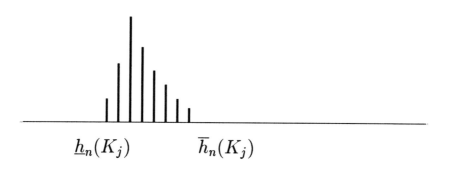

Figure 7.2: *Element of a fuzzy histogram*

Exercises

1. Calculate the characterizing functions of the fuzzy height of the non-precise histogram for the non-precise data given by the characterizing functions in figure 7.3. The classes for the histogram are also given in this figure.

2. Draw the fuzzy histogram for the data from exercise 1.

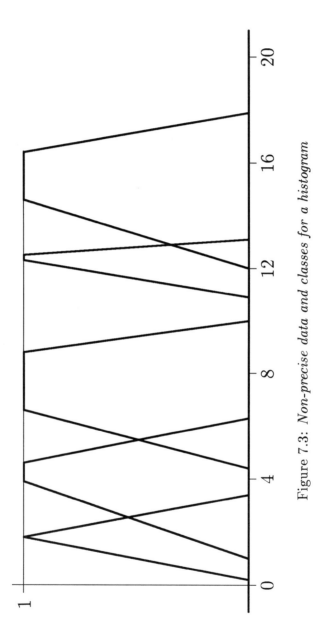

Figure 7.3: Non-precise data and classes for a histogram

8 Cumulative sums for non-precise data

For precise data in form of real numbers cumulative sums are usually taken in form of polygons using histograms. In case of non-precise data in form of non-precise numbers

$$x_1^\star, \ldots, x_n^\star$$

with corresponding characterizing functions

$$\xi_1(\cdot), \ldots, \xi_n(\cdot)$$

a reasonable *generalization of the cumulative sum* is the function

$$S_n(x) := \frac{\displaystyle\sum_{i=1}^{n} \int_{-\infty}^{x} \xi_i(t)dt}{\displaystyle\sum_{i=1}^{n} \int_{-\infty}^{\infty} \xi_i(t)dt}.$$

In figure 8.1, six non-precise observations and the corresponding cumulative sum function are depicted.

Remark 8.1: Cumulative sums are defined only if all observations are non-precise. If there are exact observations x_i in the sample $x_1^\star, \ldots, x_n^\star$ for some i, then the smoothed empirical distribution function from section 16.1 should be used.

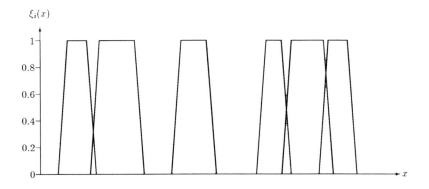

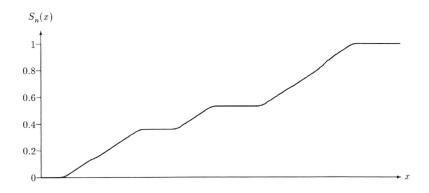

Figure 8.1: *Cumulative sum function for non-precise observations*

Exercises

1. What kind of curves are obtained as cumulative sums if only interval data are considered?

2. Having in mind the interrelation between cumulative sums and distribution functions a smoothed empirical distribution function can be defined for non-precise data (compare section 16.1). For the data given in figure 8.2 draw $S_n(\cdot)$ and $F_n^\star(\cdot)$ from section 16.1 and show the difference.

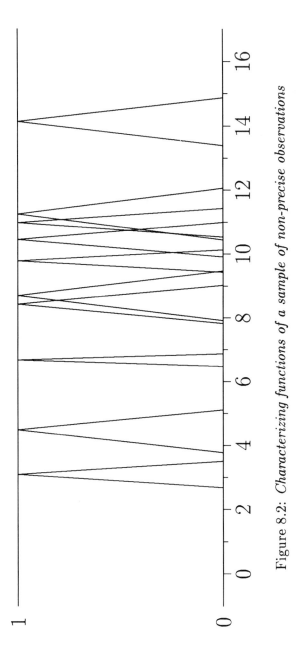

Figure 8.2: Characterizing functions of a sample of non-precise observations

9 Empirical distribution function for non-precise data

The empirical distribution function (e.d.f.) $\hat{F}_n(\cdot)$ based on n exact real observations $x_1, \ldots, x_n$ is defined by

$$\hat{F}_n(x) = \frac{1}{n} \sum_{i=1}^{n} I_{(-\infty, x]}(x_i) \qquad \forall \; x \in \mathbb{R}.$$

For non-precise observations $x_1^\star, \ldots, x_n^\star$ with characterizing functions $\xi_1(\cdot), \ldots, \xi_n(\cdot)$ for which $supp(\xi_i(\cdot))$ is finite $\forall \; i = 1(1)n$, the empirical distribution function can be generalized in different ways. Compare also section 16.

9.1 Interval data

For interval data $x_i^\star \; \hat{=} \; I_{[a_i, b_i]}(\cdot), \quad i = 1(1)n$ the generalization

$$\hat{F}_n(\cdot \mid x_1^\star, \ldots, x_n^\star)$$

of the e.d.f. is an *interval valued function* as depicted in figure 9.1.

If some observations are exact, then the generalized empirical distribution function has a step of magnitude $1/n$ at these observation points.

Remark 9.1: The shaded area in figure 9.1 covers all possible e.d.f. for exact data which could be obtained for all possible exact observations in the intervals which are recorded as non-precise data.

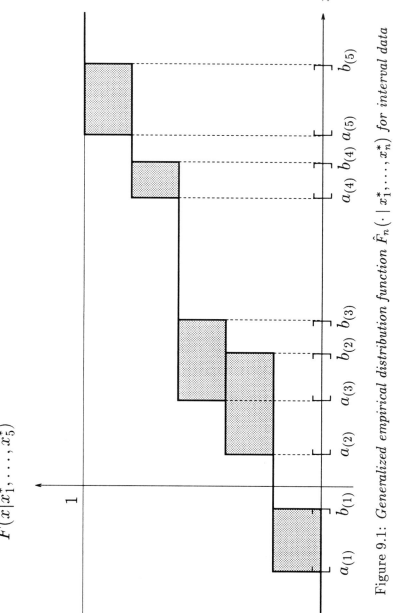

Figure 9.1: *Generalized empirical distribution function* $\hat{F}_n\left(\,\cdot\mid x_1^*,\ldots,x_n^*\right)$ *for interval data*

9.2 General non-precise data

For non-precise sample $x_1^*, \cdots, x_n^*$ with characterizing functions $\xi_1(\cdot), \cdots, \xi_n(\cdot)$ a generalization of the empirical distribution function can be graphically constructed. The construction is explained in figure 9.2. This generalization can be considered as a "mountain" over the $\big(x, F(x)\big)$-plane. The projections of the α-level curves of this mountain to the $\big(x, F(x)\big)$-plane is characteristic for this generalized empirical distribution function.

In case all characterizing functions $\xi_i(\cdot)$ are of the same shape the projections of the α-level curves of the mountain have the shape of distribution functions. An example is given in figure 9.2.

Remark 9.2: For precise observations x_i with $x_i \hat{=} x_{(k)}$ this generalization of the empirical distribution function has a jump from 0 to 1 over the interval $\{(x_i, y) : \frac{k-1}{n} < y \le \frac{k}{n}\}$ in the $\big(x, F(x)\big)$-plane.

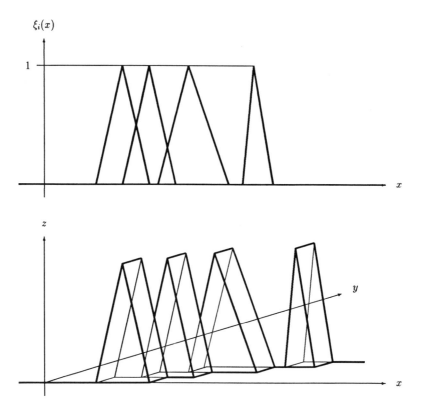

Figure 9.2: *Non-precise observations and corresponding generalized empirical distribution function*

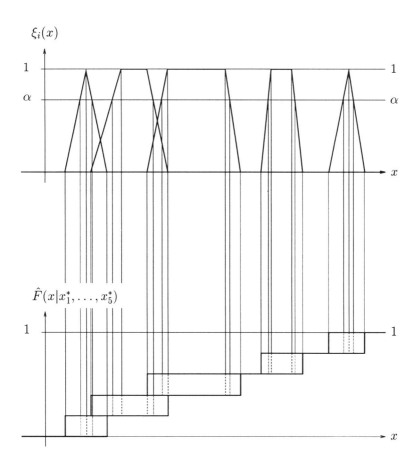

Figure 9.3: *Non-precise observations and α-level curves for the generalized e.d.f.*

Exercises

1. Draw an axonometric picture of the generalized empirical distribution function for interval data with several precise observations.

2. For a non-precise sample with characterizing functions given in figure 9.4, draw the lower and upper α-level curves of the corresponding generalized empirical distribution function for $\alpha = 0.1(0.1)1$.

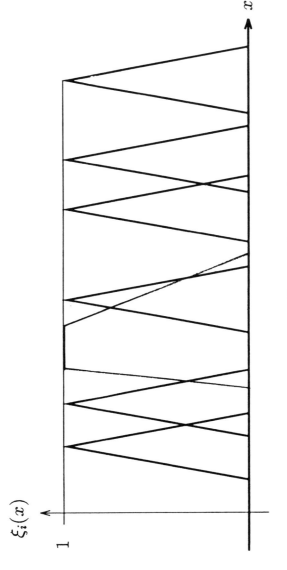

Figure 9.4: *Non-precise sample*

10 Empirical fractiles for non-precise data

For generalized empirical distribution functions and $0 < p < 1$ generalized p-fractiles can be defined in the following way.

Let

$$\hat{F}_n(\cdot \mid x_1^\star, \dots, x_n^\star)$$

be the non-precise empirical distribution function and $0 < p < 1$. Then the

non-precise p-fractile $x_p^\star$

is the non-precise number with α-cut presentation

$$\left(B_\alpha(x_p^\star); \ \alpha \in (0,1] \right)$$

of its characterizing function, given by

$$B_\alpha(x_p^\star) := \left[\bar{\hat{F}}_{n;\alpha}^{-1}(p), \ \hat{\underline{F}}_{n;\alpha}^{-1}(p) \right],$$

where

$$\hat{\underline{F}}_{n;\alpha}^{-1}(p)$$

and

$$\bar{\hat{F}}_{n;\alpha}^{-1}(p)$$

are obtained as explained in figure 10.1.

The non-precise p-fractile is the non-precise number

$$x_p^\star \text{ with characterizing function } \varphi_{x_p^\star}(\cdot)$$

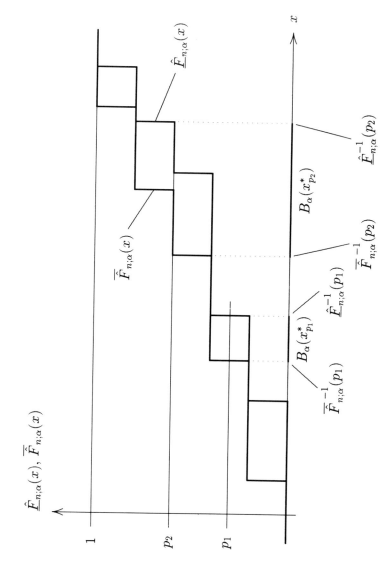

Figure 10.1: *Construction of α-cuts of non-precise fractiles*

given by its α-cuts

$$B_\alpha(x_p^\star) \qquad \forall \quad \alpha \in (0,1].$$

Therefore using proposition 2.1, $\varphi_{x_p^\star}(\cdot)$ is given by

$$\varphi_{x_p^\star}(x) = \max_{\alpha \in (0,1]} \alpha \cdot I_{B_\alpha}(x) \qquad \forall \quad x \in \mathbb{R}.$$

Exercises

1. Determine the α-cuts of the non-precise p-fractiles $x_p^\star$ of the generalized e.d.f. for the data given in exercise 1 of section 9, for $\alpha = 0.1(0.1)1$.

2. What are the α-cuts of the non-precise p-fractiles for interval data?

Chapter III

Foundations for statistical inference with non-precise data

11 Combination of non-precise observations

Considering functions

$$t(x_1, \ldots, x_n)$$

of precise observations $x_1, \ldots, x_n$, so-called *statistics*, it is necessary to generalize this concept for the situation of non-precise observations $x_1^\star, \ldots, x_n^\star$ with corresponding characterizing functions $\xi_1(\cdot), \ldots, \xi_n(\cdot)$.

For precise observations $x_1, \ldots, x_n$ with $x_i \in M$, where M is the *observation space*, statistics $t(x_1, \ldots, x_n)$ are measurable functions from the *sample space* $M^n = M \times \ldots \times M$ into another measurable space, for example $\mathbb{R}$.

11.1 One-dimensional quantities

Examples for $t(x_1, \ldots, x_n)$ are the *sample mean* $\overline{x}_n$

$$\overline{x}_n = t_1(x_1, \ldots, x_n) := \frac{1}{n} \sum_{i=1}^{n} x_i$$

and the *sample variance* s_n^2

$$s_n^2 = t_2(x_1, \ldots, x_n) := \frac{1}{n-1} \sum_{i=1}^{n} (x_i - \overline{x}_n)^2.$$

Remark 11.1: For the above-mentioned functions the precise observations $x_1, \ldots, x_n$, with $x_i \in M$, are combined to a vector $\underline{x} = (x_1, \ldots, x_n)$ which is an element of the sample space M^n.

The generalization of statistical functions to the situation of non-precise data makes it necessary to combine the non-precise elements of the observation space M to a non-precise element of the sample space M^n in order to allow the application of definition 5.2 and proposition 5.1.

Combining n non-precise observations

$$x_1^\star, \ldots, x_n^\star$$

with corresponding characterizing functions

$$\xi_1(\cdot), \ldots, \xi_n(\cdot)$$

should be done reasonably in such a way that the resulting characterizing function

$$\xi : M^n \to [0, 1]$$

of the *non-precise combined element* $\underline{x}^\star$ of the sample space M^n is a non-precise vector in the sense of definition 4.1 from section 4.

In general, reasonable *combination-rules* $K_n(\cdot, \cdots, \cdot)$ in

$$\xi(x_1, \ldots, x_n) := K_n\big(\xi_1(x_1), \ldots, \xi_n(x_n)\big)$$

have to fulfill the following properties.

(1) $K_1\big(\xi_1(x)\big) = \xi_1(x)$

(2) For all fixed real numbers $\overset{\circ}{x}_1, \ldots, \overset{\circ}{x}_n$

$$K_n\left(I_{\{\overset{\circ}{x}_1\}}(x_1), \ldots, I_{\{\overset{\circ}{x}_n\}}(x_n)\right) = I_{\{(\overset{\circ}{x}_1, \ldots, \overset{\circ}{x}_n)\}}(x_1, \ldots, x_n)$$

(3) $K_n\big(I_{[a_1,b_1]}(x_1), \ldots, I_{[a_n,b_n]}(x_n)\big)$
$$= I_{[a_1,b_1] \times \ldots \times [a_n,b_n]}(x_1, \ldots, x_n).$$

(4) For characterizing functions $\xi_1(\cdot), \ldots, \xi_n(\cdot)$ the resulting $\xi(\cdot, \cdots, \cdot)$ must be a characterizing function of a non-precise vector.

The following combination rules are used:

Minimum combination-rule (short: *Minimum-rule*)

$$\xi(x_1,\ldots,x_n) := \min_{i=1(1)n} \xi_i(x_i)$$

Product combination-rule: (short: *Product-rule*)

$$\xi(x_1,\ldots,x_n) := \prod_{i=1}^{n} \xi_i(x_i)$$

Remark 11.2: The minimum-rule is motivated by the so-called extension principle from fuzzy set theory. Moreover computations are easier using the minimum-rule and the α-cuts of the non-precise combined sample element are easy to obtain from the α-cuts of the non-precise observations.

The advantage of the product-rule is a kind of consistency of estimators for increasing sample size n.

Lemma 11.1: Let $x_1^\star,\ldots,x_n^\star$ be a sample of non-precise observations with characterizing functions $\xi_1(\cdot),\ldots,\xi_n(\cdot)$. If the minimum-rule is used to obtain the non-precise combined sample element $\underline{x}^\star$, then the α-cuts of $\underline{x}^\star$ are the Cartesian products of the α-cuts of the non-precise observations $x_i^\star$, i.e.,

$$B_\alpha(\underline{x}^\star) = X_{i=1}^n B_\alpha(x_i^\star) \qquad \forall \quad \alpha \in (0,1].$$

Proof: For all $\alpha \in (0,1]$ we have

$$
\begin{aligned}
B_\alpha(\underline{x}^\star) &= \{\underline{x} \in \mathbb{R}^n : \xi(\underline{x}) \geq \alpha\} \\
&= \{\underline{x} \in \mathbb{R}^n : \min_{i=1(1)n} \xi_i(x_i) \geq \alpha\} \\
&= \{\underline{x} \in \mathbb{R}^n : \xi_i(x_i) \geq \alpha \quad \forall\, i = 1(1)n\} \\
&= X_{i=1}^n \{x_i \in \mathbb{R} : \xi_i(x_i) \geq \alpha\} = X_{i=1}^n B_\alpha(x_i^\star).
\end{aligned}
$$

$\square$

Lemma 11.2: Let $x_1^\star,\ldots,x_n^\star$ be non-precise numbers with characterizing functions $\xi_1(\cdot),\ldots,\xi_n(\cdot)$, then the functions

$$\xi(x_1,\ldots,x_n) := \min_{i=1(1)n} \xi_i(x_i)$$

and

$$\varphi(x_1, \ldots, x_n) := \prod_{i=1}^{n} \xi_i(x_i)$$

are characterizing functions of non-precise vectors in the sense of definition 4.1.

Proof: For both functions $\xi(\cdot)$ and $\varphi(\cdot)$ $\exists$ $\underline{x}_0 \in M^n$ with $\xi(\underline{x}_0) = 1$ and $\exists$ $\underline{x}_1 \in M^n$ with $\varphi(\underline{x}_1) = 1$. Moreover we have to prove that the α-cuts $B_\alpha(\xi(\cdot))$ and $B_\alpha(\varphi(\cdot))$ are simply connected and compact subsets of $\mathbb{R}^n$. Generally α-cuts of non-precise combined sample elements obtained by the product-rule are subsets of the α-cuts of the non-precise combined sample element obtained by the minimum-rule. Therefore $B_\alpha(\varphi(\cdot)) \subseteq B_\alpha(\xi(\cdot))$ $\forall$ $\alpha \in (0, 1]$. By lemma 11.1 we have

$$B_\alpha(\xi(\cdot)) = \mathrm{X}_{i=1}^{n} B_\alpha(\xi_i(\cdot))$$

and therefore $B_\alpha(\xi(\cdot))$ is closed and bounded and also simply connected.

$\square$

The combination of non-precise observations $x_1^\star, \ldots, x_n^\star$ yields a non-precise element $\underline{x}^\star$ of the sample space which is described by its characterizing function. For two one-dimensional observations $x_1^\star$ and $x_2^\star$ with corresponding characterizing functions $\xi_1(\cdot)$ and $\xi_2(\cdot)$ this is depicted in figure 11.1.

Using the product-rule the α-cuts of the non-precise combined sample element can be explicitly calculated for special forms of non-precise data.

Proposition 11.1: Let $x_i^\star = e^\star(m_i, m_i, a_i, a_i, p, p)$, $i = 1(1)n$ be n non-precise observations whose characterizing functions are of exponential type as explained in section 2. For the product rule the α-cuts $B_\alpha(\underline{x}^\star)$ of the non-precise combined sample element $\underline{x}^\star$ are linear transformed closed unit spheres relative to the corresponding p-norm, i.e.,

$$B_\alpha(\underline{x}^\star) = \left\{ \underline{x} : \sum_{i=1}^{n} \left| \frac{x_i - m_i}{a_i(-\ln \alpha)^{1/p}} \right|^p \leq 1 \right\}$$

with $\underline{x} = (x_1, \ldots, x_n) \in \mathbb{R}^n$.

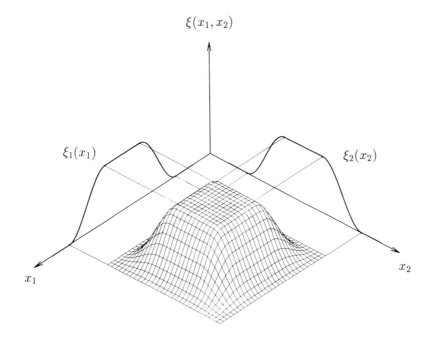

$$\xi(x_1, x_2)$$

$$\xi_1(x_1) \qquad\qquad \xi_2(x_2)$$

$$x_1 \qquad\qquad x_2$$

$\xi_1(\cdot)$ characterizing function of $x_1^\star$
$\xi_2(\cdot)$ characterizing function of $x_2^\star$
$\xi(\cdot, \cdot)$ characterizing function of $\underline{x}^\star$

Figure 11.1: *Characterizing function of a non-precise combined sample element for two observations*

Proof: For the α-cut $B_\alpha(\underline{x}^\star)$ with $\varphi(\cdot) \hat{=} \underline{x}^\star$ we have the equivalence

$$\varphi(\underline{x}) \geq \alpha \iff \prod_{i=1}^{n} e^{-|\frac{x_i - m_i}{a_i}|^p} \geq \alpha$$

$$\Leftrightarrow -\sum_{i=1}^{n} | \frac{x_i - m_i}{a_i} |^p \geq \ln \alpha$$

$$\Leftrightarrow \sum_{i=1}^{n} |\frac{x_i - m_i}{a_i(-\ln \alpha)^{1/p}} |^p \leq 1.$$

$$\square$$

Remark 11.3 : For $p = 2$ the Euclidian norm is obtained. For $n = 2$ and $p = 1$ the α-cuts are squares, i.e., the unit sphere relative to this norm. For $p = 2$ the α-cuts are ellipses, i.e., linear transformed circles. An example is given in figure 11.2.

11.2   Vector-valued quantities

In taking non-precise observations from k-dimensional vector-valued stochastic quantities the characterizing functions of the observations are assumed to be functions as described in definition 4.1. For n non-precise observations $\underline{x}_1^\star, \ldots, \underline{x}_n^\star$ of such quantities the data are given by n characterizing functions

$$\xi_i(x_1, \ldots, x_k) \quad \text{for } i = 1(1)n$$

of k real variables.

Here the observation space is a subset of $\mathbb{R}^k$, i.e., $M_X \subseteq \mathbb{R}^k$, and the non-precise combined sample element is a non-precise element of the sample space $M_X^n \subseteq \mathbb{R}^{n \cdot k}$.

The characterizing function $\xi(\cdot, \cdots, \cdot)$ of the non-precise combined sample element is also obtained by a more general combination rule $K_n(\cdot, \cdots, \cdot)$ which is now a function of the n functions $\xi_i(\cdot, \cdots, \cdot)$ which are real functions of k real variables $x_1, \ldots, x_k$:

$$\xi(x_1, x_2, \ldots, x_{k \cdot n}) = K_n \Big(\xi_1(x_1, \ldots, x_k), \; \xi_2(x_{k+1}, \ldots, x_{2k}),$$

$$\xi_3(x_{2k+1}, \ldots, x_{3k}), \ldots, \; \xi_n(x_{(n-1)k+1}, \ldots, x_{k \cdot n}) \Big).$$

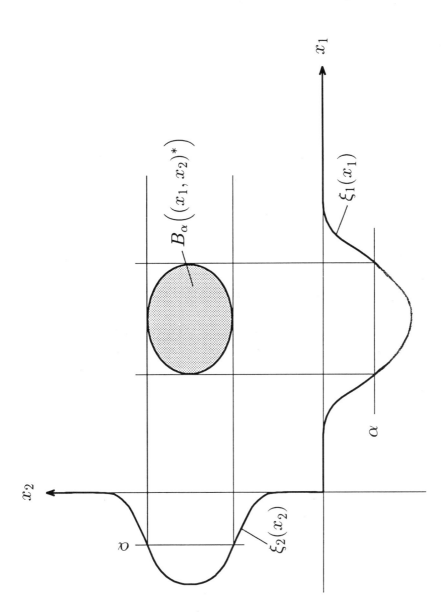

Figure 11.2: α-Cuts for the product-rule for two observations with exponential characterizing functions

The most used combination-rule in this situation is the *generalized minimum combination-rule*, i.e.,

$$\xi(x_1,\ldots,x_{k\cdot n}) = \min_{i=1(1)n} \ \xi_i(x_{(i-1)k+1}, x_{(i-1)k+2}, \ldots, x_{(i-1)k+k}).$$

For two-dimensional non-precise observations we obtain characterizing functions of the form

$$\xi_i(x,y) \quad \text{for} \quad i = 1(1)n$$

and characterizing function $\xi(x_1, y_1, x_2, y_2, \ldots, x_n, y_n)$ of the non-precise combined sample element by

$$\xi(x_1, y_1, \ldots, x_n, y_n) = \min_{i=1(1)n} \ \xi_i(x_i, y_i).$$

Remark 11.4: It is also possible to use other combination-rules, for example a generalized product-rule. But for applications the minimum combination-rule is suitable and the necessary calculations are also practicable.

An application of this concept is explained in section 16.4 where the empirical correlation coefficient is generalized to the situation of non-precise observations of a two-dimensional stochastic quantity.

Exercises

1. Let $\xi_1(\cdot), \ldots, \xi_n(\cdot)$ be the characterizing functions of n non-precise observations, and

$$\xi(x_1, \ldots, x_n) = \min_{i=1(1)n} \xi_i(x_i)$$

$$\varphi(x_1, \ldots, x_n) = \prod_{i=1}^{n} \xi_i(x_i).$$

Show that for the α-cuts of $\xi(\cdot, \ldots, \cdot)$ and $\varphi(\cdot, \ldots, \cdot)$ the following holds

$$B_\alpha\big(\varphi(\cdot)\big) \subseteq B_\alpha\big(\xi(\cdot)\big) \quad \text{for all} \ \ \alpha \geq 0.$$

2. Draw an axonometric picture of the characterizing function of the non-precise combined sample element for two observations with characterizing functions $\xi_1(\cdot)$ and $\xi_2(\cdot)$ given by the following diagrams:

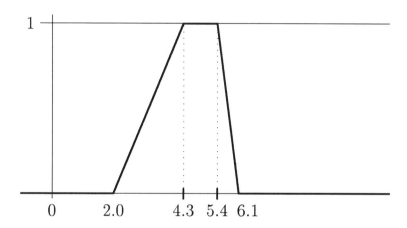

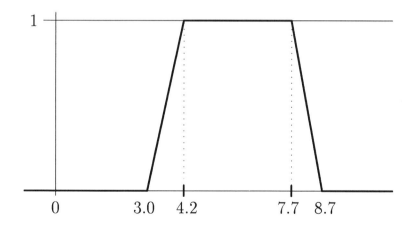

Is there a difference for the minimum combination-rule and the product combination-rule?

12   Sample moments for non-precise observations

For precise samples $x_1, \ldots, x_n$ of a stochastic quantity the sample mean

$$\overline{x}_n = \frac{1}{n} \sum_{i=1}^{n} x_i$$

and the statistic

$$\frac{1}{n} \sum_{i=1}^{n} (x_i - \overline{x}_n)^2$$

are special cases of so-called *sample moments*

$$m'_r := \frac{1}{n} \sum_{i=1}^{n} x_i^r$$

and

$$m_r := \frac{1}{n} \sum_{i=1}^{n} (x_i - \overline{x}_n)^r \quad \text{for } r \in \mathbb{N}$$

respectively.

For non-precise data $x_1^\star, \ldots, x_n^\star$ with characterizing functions $\xi_1(\cdot), \ldots, \xi_n(\cdot)$ the concept of section 11 can be applied.

Let $\xi(\cdot, \ldots, \cdot)$ be the characterizing function of the non-precise combined sample element $\underline{x}^\star$. Then the generalized *non-precise sample moment* $m_r^\star$ is given as a non-precise number with characterizing function $\varphi(\cdot)$ given by definition 5.2 from section 5 in the following way:

$$\varphi(y) = \sup_{x_i \in \mathbb{R}} \left\{ \xi(x_1, \ldots, x_n) : \frac{1}{n} \sum_{i=1}^{n} (x_i - \overline{x}_n)^r = y \right\}.$$

Proposition 12.1: The α-cuts $B_\alpha(m_r^\star)$ for $\alpha \in (0, 1]$ are given by application of proposition 5.1 as follows:

$$B_\alpha(m_r^\star) = \left[\min_{\underline{x} \in B_\alpha(\underline{x}^\star)} \frac{1}{n} \sum_{i=1}^{n} (x_i - \overline{x}_n)^r, \quad \max_{\underline{x} \in B_\alpha(\underline{x}^\star)} \frac{1}{n} \sum_{i=1}^{n} (x_i - \overline{x}_n)^r \right],$$

with $\underline{x} = (x_1, \ldots, x_n) \in \mathbb{R}^n$, and $B_\alpha(\underline{x}^\star) = B_\alpha\big(\xi(\cdot, \ldots, \cdot)\big)$ is the α-cut of the non-precise combined sample element.

Proof: The conclusion follows from the continuity of the function

$$\sum_{i=1}^{n} (x_i - \overline{x}_n)^r$$

and proposition 5.1.

$\square$

12.1   Sample mean for non-precise data

The sample mean $\overline{x}_n^\star$ for non-precise observations $x_1^\star, \ldots, x_n^\star$ is the special case of the sample moment m_r' for $r = 1$.

If $\xi(\cdot, \ldots, \cdot)$ is the characterizing function of the non-precise combined sample element $\underline{x}^\star$ then the characterizing function $\varphi_{\overline{x}_n^\star}(\cdot)$ of the *non-precise sample mean* $\overline{x}_n^\star$ is given by its values

$$\varphi_{\overline{x}_n^\star}(x) = \sup\{\xi(x_1, \ldots, x_n) : \overline{x}_n = x\}$$

and for the α-cuts $B_\alpha(\overline{x}_n^\star)$ we obtain

$$B_\alpha(\overline{x}_n^\star) = \left[\min_{\underline{x} \in B_\alpha(\underline{x}^\star)} \overline{x}_n, \quad \max_{\underline{x} \in B_\alpha(\underline{x}^\star)} \overline{x}_n \right],$$

where $B_\alpha(\underline{x}^\star)$ are the α-cuts of the non-precise combined sample $\underline{x}^\star$.

Examples of non-precise sample means are given in figure 12.1 and figure 12.2. The minimum-rule is applied for figure 12.1.

Application of the product-rule for combination of the same non-precise observations yields the non-precise sample mean given in figure 12.2.

Remark 12.1: If non-precise samples $x_1^\star, \ldots, x_n^\star$ are combined by the minimum-rule then the α-cuts $B_\alpha(\bar{x}_n^\star)$ of the non-precise sample mean $\bar{x}_n^\star$ are easy to obtain from the α-cuts $B_\alpha(x_i^\star)$ of the non-precise observations $x_i^\star$, i.e.,

$$B_\alpha(\bar{x}_n^\star) = \left[\frac{1}{n} \sum_{i=1}^n \underline{B}_\alpha(x_i^\star), \ \frac{1}{n} \sum_{i=1}^n \overline{B}_\alpha(x_i^\star) \right] \qquad \forall \ \alpha \in (0,1],$$

where the α-cuts of the i-th non-precise observation $x_i^\star$ are the following intervals

$$B_\alpha(x_i^\star) = \left[\underline{B}_\alpha(x_i^\star), \ \overline{B}_\alpha(x_i^\star) \right] \qquad \forall \ \alpha \in (0,1].$$

Proof: For the minimum-rule to obtain the non-precise combined sample element $\underline{x}^\star$ we have by lemma 11.1

$$B_\alpha(\underline{x}^\star) = B_\alpha(x_1^\star) \times \ldots \times B_\alpha(x_n^\star).$$

The above form of the α-cuts $B_\alpha(\bar{x}_n^\star)$ of the sample mean $\bar{x}_n^\star$ follows from the fact that the function

$$\sum_{i=1}^n x_i \text{ is monotone increasing in all variables } x_i$$

and assumes its minimum for

$$\left(\underline{B}_\alpha(x_1^\star), \ldots, \underline{B}_\alpha(x_n^\star) \right).$$

A similar argument is valid for the maximum.

$\square$

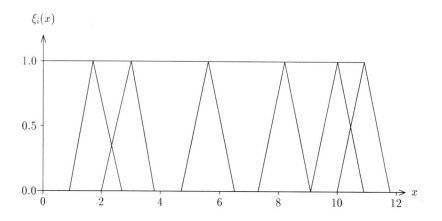

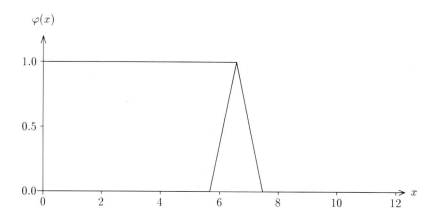

$\psi(\cdot)$ characterizing function of $\overline{x}_6^\star$ using the minimum-rule.

Figure 12.1: *Non-precise sample mean using the minimum-rule*

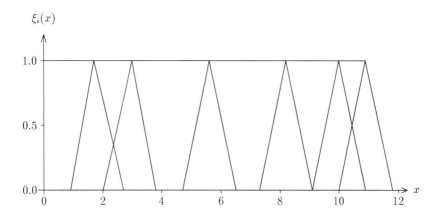

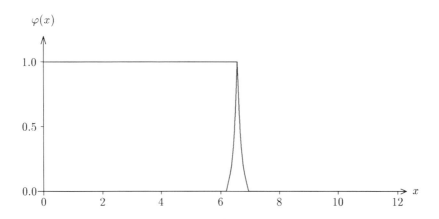

$\varphi(\cdot)$ characterizing function of $\overline{x}_6^\star$ using the product-rule.

Figure 12.2: *Non-precise sample mean using the product rule*

12.2   Sample variance for non-precise data

The generalization of the sample variance

$$s_n^2 = \frac{1}{n-1} \sum_{i=1}^{n} (x_i - \overline{x}_n)^2$$

to the situation of non-precise data $x_1^\star, \ldots, x_n^\star$ with characterizing functions

$$\xi_1(\cdot), \ldots, \xi_n(\cdot)$$

is using the non-precise combined sample element $\underline{x}^\star$ with corresponding characterizing function $\xi(\cdot, \ldots, \cdot)$ via proposition 5.1.

The characterizing function $\psi(\cdot)$ of the *non-precise sample variance* $(s_n^2)^\star$ is given by its values

$$\psi(y) = \sup \left\{ \xi(x_1, \ldots, x_n) : \quad \frac{1}{n-1} \sum_{i=1}^{n} (x_i - \overline{x}_n)^2 = y \right\}$$

for all $y \geq 0$.

Using the following notations

$$\underline{x} = (x_1, \ldots, x_n)$$

and

$$g(\underline{x}) = g(x_1, \ldots, x_n) = \frac{1}{n-1} \sum_{i=1}^{n} (x_i - \overline{x}_n)^2$$

and

$$B_\alpha(\underline{x}^\star)$$

for the α-cut of the non-precise combined sample element $\underline{x}^\star$ the α-cuts

$$B_\alpha \left((s_n^2)^\star \right)$$

of the non-precise sample variance $(s_n^2)^\star$ are given by

$$B_\alpha \left((s_n^2)^\star \right) = \left[\min_{\underline{x} \in B_\alpha(\underline{x}^\star)} g(\underline{x}), \quad \max_{\underline{x} \in B_\alpha(\underline{x}^\star)} g(\underline{x}) \right] \qquad \forall \ \alpha \in (0,1].$$

In an analogous way a non-precise generalization $s_n^\star$ of the sample dispersion

$$s_n = +\sqrt{s_n^2}$$

is obtained.

In figure 12.3 a non-precise sample and the corresponding non-precise sample variance as well as the non-precise sample dispersion is depicted using the minimum rule.

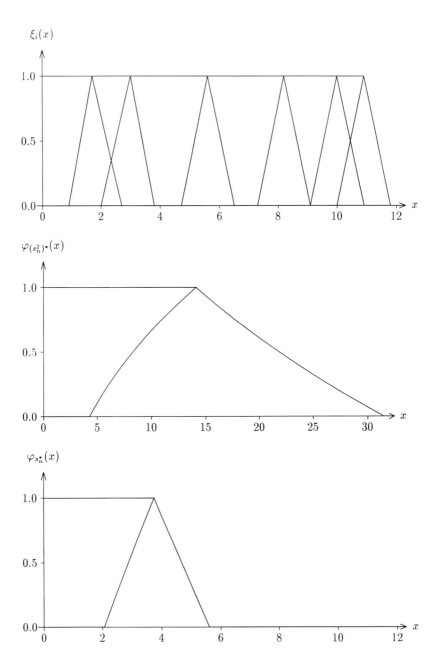

Figure 12.3: *Non-precise sample variance $(s_n^2)^\star$ and non-precise sample dispersion $s_n^\star$ using the minimum-rule*

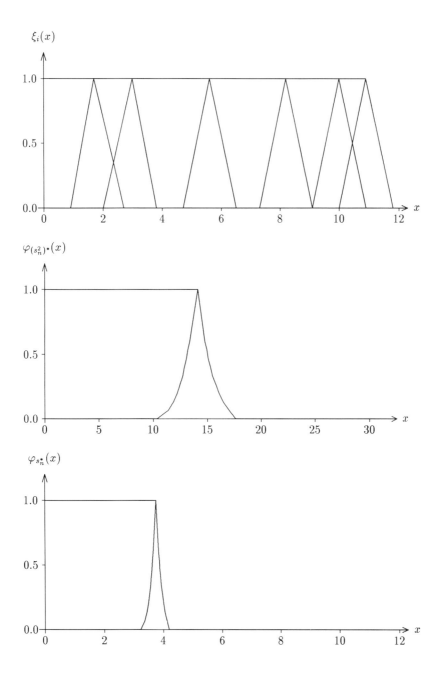

Figure 12.4: *Non-precise sample variance and non-precise sample dispersion using the product rule*

Exercises

1. Take the sample from figure 12.1 and calculate the characterizing function of the non-precise sample mean for only four non-precise obervations using the minimum-rule. Compare the imprecision of the result with the imprecision of the non-precise sample mean for all observations.

2. Study the relationship between the imprecision of a positive non-precise number $x^\star$ with characterizing function $\xi(\cdot)$ and the imprecision of the corresponding non-precise number $+\sqrt{x^\star}$.

13 Sequences of non-precise observations

In statistical inference for precise observations, limit theorems such as the law of large numbers are important. It is possible to generalize the law of large numbers to the situation of sequences $x_1^\star, x_2^\star, \ldots$ of non-precise observations of a stochastic quantity X. There are also generalizations of the central limit theorem (compare [14]).

13.1 Generalization of the law of large numbers

Theorem 13.1 (Bernoulli's law of large numbers): For a sequence of independent observations of a stochastic quantity X and a fixed event A with $p_A = $ Probability $\{X \in A\}$ and two arbitrary positive real numbers ϵ and δ there exists a number $N(\epsilon, \delta) \in \mathbb{N}$ such that for all $n > N(\epsilon, \delta)$ and relative frequency $h_n(A)$

$$Probability \ \{|\ h_n(A) - p_A\ | < \epsilon\} \geq 1 - \delta \ .$$

The imprecision of observations generates imprecision of the relative frequencies as explained in section 7. A more detailed discussion is given in [17] of the references. In accordance with the mentioned reference, the *non-precise relative frequency* of an event A after n observations is denoted by ν_A in this section.

ν_A is a non-precise element of the ordered set

$$\left\{ \frac{i}{n} \ : \ i \in \{0, 1, \ldots, n\} \right\}$$

with characterizing function $\varphi_{\nu_A}(\cdot)$.

For non-precise rational numbers ν with characterizing function $\varphi_\nu(\cdot)$ we define

$$\mathcal{L}_\nu = \{\varphi_\nu(\frac{i}{n}) > 0 : \quad \frac{i}{n} \leq C_1^L(\nu)\}$$

$$\mathcal{R}_\nu = \{\varphi_\nu(\frac{i}{n}) > 0 : \quad \frac{i}{n} \geq C_1^U(\nu)\}$$

with

$$\left. \begin{array}{l} C_\alpha^L(\nu) = \min\{\frac{i}{n} : \quad \varphi_\nu(\frac{i}{n}) \geq \alpha\} \\ C_\alpha^U(\nu) = \max\{\frac{i}{n} : \quad \varphi_\nu(\frac{1}{n}) \geq \alpha\} \end{array} \right\} \alpha \in (0,1]$$

and for every non-precise rational number ν and $p \in \mathcal{F}([0,1])$ we define

$$\|\nu - p\|_\nu = \max \left\{ \max_{\alpha \in \mathcal{L}_\nu} |C_\alpha^L(\nu) - C_\alpha^L(p)|, \max_{\alpha \in \mathcal{R}_\nu} |C_\alpha^U(\nu) - C_\alpha^U(p)| \right\}.$$

Using the above concepts a generalization of Bernoulli's law of large numbers for non-precise observations can be proved.

Theorem 13.2 (Generalized Bernoulli's law of large numbers for non-precise observations): Under the conditions of theorem 13.1 given a sequence of non-precise observations of X there exists a non-precise number $p \in \mathcal{F}([0,1])$ with $\varphi_p(p_A) = 1$ and for arbitrary small $\epsilon > 0$ and $\delta > 0$ there exists a number $N(\epsilon, \delta) \in \mathbb{N}$ with

$$Probability \, \{\|\nu_A(n) - p\|_{\nu_A(n)} < \varepsilon\} \geq 1 - \delta$$

for all $n > N(\epsilon, \delta)$ if the characterizing functions of the observations are continuous.

The proof is given in reference [17] of the list of references.

Remark 13.1: There are also other generalizations of the law of large numbers. Compare reference [14] and references given therein.

13.2   Convergence of functions of non-precise observations

Considering functions

$$s(x_1^\star, \ldots, x_n^\star)$$

of non-precise observations $x_1^\star, \ldots, x_n^\star$, we obtain the characterizing function of the non-precise value $s(x_1^\star, \ldots, x_n^\star)$ by definition 5.2.

For increasing number n of observations the question arises how the imprecision of the result behaves.

Remark 13.2: If the product-rule is used to obtain the non-precise combined sample element $\underline{x}^\star$, then in general the imprecision of the non-precise values of functions is decreasing with increasing number n of observations. This is a consequence of the property of characterizing functions to have values in $[0,1]$. By multiplication of characterizing functions the values of $\varphi_{\underline{x}^\star}(\cdot)$ decrease with increasing n.

Remark 13.3: In situations where a quantity is observed for which a true value exists but observations are non-precise, it can be reasonable to use the product-rule. In this case a kind of consistency to the true value is obtained for n tending to infinity.

If no true value exists it may be reasonable to use the minimum-rule, because then the imprecision of the results does not decrease.

Exercises

1. Consider a series $x_1^\star, x_2^\star, \ldots$ of triangular non-precise numbers with identical shape. For different values of n use the product combination-rule and calculate the characterizing function $\varphi(\cdot)$ of the non-precise value of the function

$$s(x_1, \ldots, x_n) = \frac{1}{n} \sum_{i=1}^{n} x_i.$$

2. Compare the results of exercise 1 with the result which is obtained using the minimum combination-rule.

Chapter IV

Classical statistical inference for non-precise data

14   Point estimators for parameters

In this section a generalization of point estimators for statistical parameters θ of stochastic models $X \sim F_\theta$, $\theta \in \Theta$ with parameter space Θ and observation space M_X for X based on non-precise observations is explained.

An estimator $\vartheta(\cdot, \ldots, \cdot)$ for the parameter θ is a measurable function from the sample space M_X^n to Θ, i.e.,

$$\vartheta : M_X^n \to \Theta.$$

For functions $\tau(\theta)$ of the parameter θ with

$$\tau : \Theta \to \Xi = \{\tau(\theta) : \theta \in \Theta\}$$

generalizations of estimators to the situation of non-precise data can also be given. In this case estimators $t(\cdot, \ldots, \cdot)$ are functions

$$t : M_X^n \to \Xi.$$

Example 14.1: For normal distribution

$$X \sim N(\mu, \sigma^2)$$

we have
$$\theta = (\mu, \sigma^2).$$

Here an important function $\tau(\theta)$ of the parameter θ to be estimated is
$$\tau(\theta) = \mu$$

with $\Theta = \{(\mu, \sigma^2) : \mu \in \mathbb{R}, \sigma^2 > 0\}$ and $\Xi = \mathbb{R}$.

Note: The construction principles for reasonable estimators in case of precise observations are assumed to be known. These are given in all good books on statistical inference.

Let $t(X_1, \ldots, X_n)$ be an estimator for a transformed parameter $\tau(\theta) \in \mathbb{R}$ based on a sample $X_1, \ldots, X_n$ of a stochastic quantity X. For an observed sample $x_1, \ldots, x_n \in M_X^n$ an *estimated value*

$$\widehat{\tau(\theta)} = t(x_1, \ldots, x_n) \in \Xi$$

is obtained.

For *non-precise observations* $x_1^\star, \ldots, x_n^\star$ a reasonable generalization of estimators must yield a *non-precise estimated value* $\widehat{\tau(\theta)}^\star$ for $\tau(\theta)$.

The construction of the non-precise estimation uses the characterizing functions $\xi_i(\cdot)$ of the observations $x_i^\star$, $i = 1(1)n$ combined by a suitable combination-rule to obtain the non-precise combined sample element $\underline{x}^\star$ of the sample space M_X^n. This non-precise combined sample element is a non-precise vector with characterizing function $\xi(\cdot, \ldots, \cdot)$ given by its values

$$\xi(x_1, \ldots, x_n) = K_n\big(\xi_1(x_1), \ldots, \xi_n(x_n)\big) \quad \text{with } x_i \in M_X$$

as explained in section 11.

The non-precise combined sample element is the basis for the construction of a non-precise generalization of estimators for θ or functions $\tau(\theta)$.

Definition 14.1: Let $\vartheta(X_1, \ldots, X_n)$ be an estimator for the parameter θ of a stochastic model $X \sim f(. \mid \theta)$, $\theta \in \Theta$ based on a sample $X_1, \ldots, X_n$ of X. Then for non-precise observations $x_1^\star, \ldots, x_n^\star$ a non-precise estimation $\hat{\theta}^\star$ for θ based on the non-precise combined sample element $\underline{x}^\star$, whose characterizing function is $\xi(\cdot, \ldots, \cdot)$, is given by a non-precise element $\hat{\theta}^\star$ of the parameter space with characterizing function $\psi(\cdot)$, given by its values

$$\psi(\theta) := \sup \Big\{ \xi(x_1, \ldots, x_n) : \vartheta(x_1, \ldots, x_n) = \theta \Big\}.$$

To find the supremum all elements $(x_1, \ldots, x_n)$ of the sample space have to be considered for which the condition is fulfilled.

Using the notation $\underline{x} = (x_1, \ldots, x_n)$ we can write

$$\psi(\theta) = \sup_{\underline{x} \in M_X^n} \left\{ \xi(\underline{x}) : \vartheta(\underline{x}) = \theta \right\}.$$

Remark 14.1: The construction of the characterizing function of the non-precise estimation is an application of the generalization of classical functions given in section 5.

In the sample there can be also precise observations x_i. In this case the corresponding characterizing function is $I_{\{x_i\}}(\cdot)$.

In figure 14.1 the principle is explained for a sample of $n = 2$ non-precise observations. Here the parameter θ is one-dimensional, i.e. $\theta \in \Theta \subseteq \mathbb{R}$. The estimator is the sample mean and therefore the set of all vectors $\underline{x} = (x, y)$ in which the supremum

$$\sup_{\underline{x} \in M_X^n} \left\{ \xi(\underline{x}) : \vartheta(\underline{x}) = \frac{x + y}{2} = \theta \right\}$$

has to be found is the thick line in the (x, y)-plane in figure 14.1.

Example 14.2: For the stochastic model $X \sim Ex_\theta$, $\theta \in (0, \infty)$, i.e., the exponential distribution with density function

$$f(x \mid \theta) = \frac{1}{\theta} \, e^{-x/\theta} I_{(0,\infty)}(x),$$

the optimal estimator for θ based on a precise sample $X_1, \ldots, X_n$ of X is the sample mean, i.e.,

$$\vartheta(X_1, \ldots, X_n) = \overline{X}_n = \frac{1}{n} \sum_{i=1}^n X_i.$$

For non-precise observations $x_1^\star, \ldots, x_n^\star$ of X with non-precise combined sample element $\underline{x}^\star$ and corresponding characterizing function $\xi(\cdot, \ldots, \cdot)$, the characterizing function $\psi(\cdot)$ of the non-precise estimate $\theta^\star$ for θ is given by

$$\psi(\theta) = \sup \left\{ \xi(x_1, \ldots, x_n) : \overline{x}_n = \theta \right\}$$

where the supremum has to be taken over the sample space $M_X^n = (0, \infty)^n$.

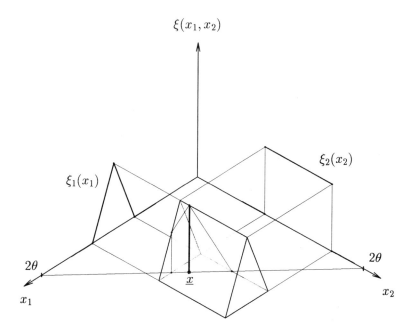

$$\hat{\theta} = \vartheta(x_1, x_2) = \frac{x_1 + x_2}{2}$$

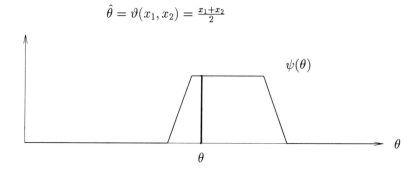

Figure 14.1: *On the construction of the characterizing function of a non-precise estimate*

In figure 14.2 a non-precise sample and the characterizing functions of non-precise estimates, using different combination rules, are depicted.

For functions $\tau(\theta)$ of parameters of stochastic models by a modification of definition 14.2 non-precise estimates can be obtained.

Definition 14.2: Under the assumptions of definition 14.1 let $t(X_1, \ldots, X_n)$ be an estimator for $\lambda = \tau(\theta)$. For non-precise observations $x_1^\star, \ldots, x_n^\star$ of X with non-precise combined sample element $\underline{x}^\star$ and corresponding characterizing function $\xi(\cdot, \ldots, \cdot)$ the non-precise estimate $\lambda^\star = \widehat{\tau(\theta)}^\star$ is given by its characterizing function $\psi(\cdot)$ with values

$$\psi(\lambda) = \sup_{\underline{x} \in M_X^n} \left\{ \xi(\underline{x}) \colon t(\underline{x}) = \lambda \right\}.$$

Example 14.3: For $X \sim N(\mu, \sigma^2)$, $\theta = (\mu, \sigma^2) \in \Theta$, $\tau(\theta) = \mu$ and non-precise sample $x_1^\star, \ldots, x_n^\star$, the characterizing function of the non-precise estimate $\hat{\mu}^\star$ is given by

$$\psi(\mu) = \sup_{\underline{x} \in \mathrm{IR}^n} \left\{ \xi(x_1, \ldots, x_n) \colon \overline{x}_n = \mu \right\}.$$

Remark 14.2: The diagrams $x \in \mathrm{IR}^n$ in this section were made using α-cuts. The basis for that is proposition 5.1.

If the minimum rule is used for the combination of observations, calculations are much simpler because it is possible to use remark 12.1.

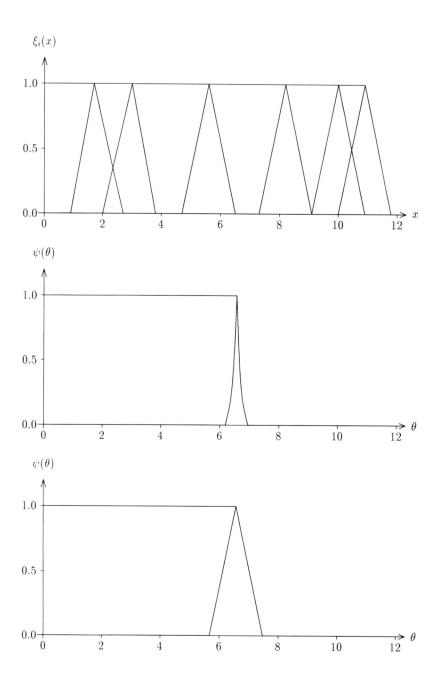

Figure 14.2: *Non-precise sample of an exponential distribu-*
tion and characterizing functions of estimators
for the mean

Exercises

1. Non-precise generalizations of point estimators are fuzzy subsets of the parameter space. Show that in case of precise data
   the definition of this section gives the indicator function of
   the exact point estimator.

2. Explain that in case of interval data for both combination-
   rules the same non-precise estimators are obtained.

15 Confidence regions for parameters

Let $\kappa(X_1,\ldots,X_n)$ be a confidence function with confidence level $1 - \alpha$ for a parameter θ, i.e., a function

$$\kappa : M_X^n \to \mathcal{P}(\Theta),$$

where $\mathcal{P}(\Theta)$ denotes the power set of the parameter space Θ, based on a sample $X_1,\ldots,X_n$ of the stochastic quantity $X \sim F_\theta$, $\theta \in \Theta$. Then $\kappa(X_1,\ldots,X_n)$ must fulfill

$$Probability \left\{ \theta \in \kappa(X_1,\ldots,X_n) \right\} = 1 - \alpha \qquad \forall \ \theta \in \Theta.$$

For observed concrete sample $x_1,\ldots,x_n$ a subset $\kappa(x_1,\ldots,x_n)$ of Θ is obtained.

In case of non-precise data $x_1^\star,\ldots,x_n^\star$ a generalization of confidence sets is obtained as a *fuzzy subset* of the parameter space in the following way.

Definition 15.1: Let $\xi(\cdot,\ldots,\cdot)$ be the characterizing function of the non-precise combined sample element and $\kappa(X_1,\ldots,X_n)$ be a confidence function. The characterizing function $\varphi(\cdot)$ of the *generalized confidence set* is given by its values

$$\varphi(\theta) := \sup \left\{ \xi(x_1,\ldots,x_n) : \theta \in \kappa(x_1,\ldots,x_n) \right\}$$

where $(x_1,\ldots,x_n)$ is varying over the sample space M_X^n of X.

Remark 15.1: For this generalized confidence set obtained by the classical confidence set $\kappa(x_1,\ldots,x_n)$ for precise data $x_1,\ldots,x_n$ the following holds

$$\varphi(\theta) = 1 \quad \text{for all} \quad \theta \in \bigcup_{(x_1,\ldots,x_n):\ \xi(x_1,\ldots,x_n)=1} \kappa(x_1,\ldots,x_n),$$

i.e., the indicator function of the union at the right hand is always below the characterizing function $\varphi(\cdot)$ of the fuzzy confidence set.

Using the abbreviation $\underline{x} = (x_1,\ldots,x_n)$ we have

$$I_{\bigcup_{\underline{x}:\ \xi(\underline{x})=1} \kappa(\underline{x})}(\theta) \leq \varphi(\theta) \quad \forall \ \theta \in \Theta.$$

This is easy to see by

$$\theta \in \bigcup_{\underline{x}:\ \xi(\underline{x})=1} \kappa(\underline{x}) \ \Rightarrow\ \exists\ \underline{x}: \xi(\underline{x}) = 1 \ \text{and}\ \theta \in \kappa(\underline{x})$$
$$\Rightarrow\ \sup\{\xi(\underline{x}): \theta \in \kappa(\underline{x})\} = 1 \ \Rightarrow\ \varphi(\theta) = 1.$$

Example 15.1: A fuzzy confidence region for the parameter $\theta = (\mu, \sigma^2)$ of a normal distribution based on non-precise data is depicted in figure 15.1.

For functions $\lambda = \tau(\theta)$ of parameters θ in stochastic models $X \sim F_\theta$, $\theta \in \Theta$ with space $\Lambda = \{\tau(\theta) : \theta \in \Theta\}$ of transformed parameter values the concept of confidence regions can be generalized also.

Definition 15.2: Let $x_1^\star,\ldots,x_n^\star$ be a non-precise sample whose non-precise combined sample element $\underline{x}^\star$ has characterizing function $\xi(\cdot,\ldots,\cdot)$. For a confidence function $\kappa(X_1,\ldots,X_n)$ for transformed parameter $\lambda = \tau(\theta)$ a *generalized confidence region* for $\lambda = \tau(\theta)$ is the fuzzy subset $\Lambda^\star$ of Λ whose characterizing function $\psi(\cdot)$ is given by its values

$$\psi(\lambda) = \sup\left\{\xi(x_1,\ldots,x_n): \lambda = \tau(\theta) \in \kappa(x_1,\ldots,x_n)\right\}$$

where for $(x_1,\ldots,x_n)$ all possible values of the sample space have to be considered.

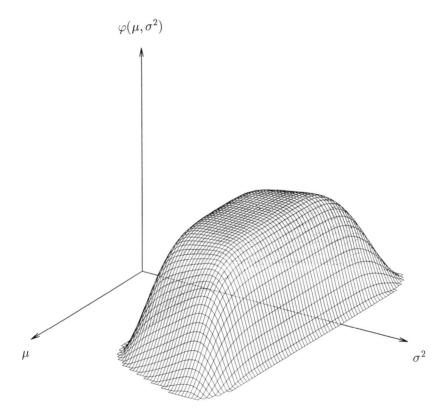

Figure 15.1: *Characterizing function of a fuzzy confidence region*

Remark 15.2: Also for transformed parameters $\lambda = \tau(\theta)$ the following relation holds:

$$I \bigcup_{\underline{x}:\ \xi(\underline{x})=1} \kappa(\underline{x})^{(\lambda)} \leq \psi(\lambda) \qquad \forall\ \lambda \in \Lambda,$$

with $\underline{x} = (x_1, \ldots, x_n)$ and $\kappa(\underline{x}) = \kappa(x_1, \ldots, x_n)$.

Example 15.2: Let X be a normally distributed stochastic quantity and $x_1^\star, \ldots, x_n^\star$ a non-precise sample of X. A generalized confidence interval for $\tau(\mu, \sigma^2) = \mu$, the expectation of $X \sim N(\mu, \sigma^2)$, should be calculated.

Let $\xi(\cdot, \ldots, \cdot)$ be the characterizing function of the non-precise combined sample element. Then the characterizing function $\psi(\cdot)$ of the generalized confidence region for μ is calculated via a classical confidence interval $\kappa(x_1, \ldots, x_n)$ for μ based on precise data $x_1, \ldots, x_n$. This is

$$\kappa(x_1, \ldots, x_n) = \left[\overline{x}_n - \frac{s_n}{\sqrt{n}} . t_{n-1;1-\frac{\delta}{2}},\ \overline{x}_n + \frac{s_n}{\sqrt{n}}\, t_{n-1,1-\frac{\delta}{2}} \right]$$

for confidence level $1 - \delta$. The characterizing function $\psi(\cdot)$ of the generalized *fuzzy confidence interval* based on the non-precise data is given — using the abbreviation $\underline{x} = (x_1, \cdots, x_n)$ — by its values

$$\psi(\mu) = \sup\left\{ \xi(\underline{x}) : \mu \in \left[\overline{x}_n - \frac{s_n}{\sqrt{n}} t_{n-1;1-\frac{\delta}{2}},\ \overline{x}_n + \frac{s_n}{\sqrt{n}} t_{n-1;1-\frac{\delta}{2}} \right] \right\}.$$

In figure 15.2 a sample of non-precise observations of a normal distribution and the corresponding fuzzy confidence interval for μ are drawn.

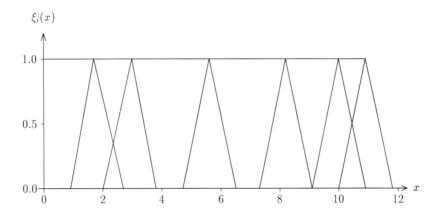

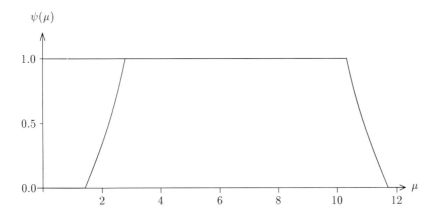

Figure 15.2: *Non-precise data and fuzzy confidence interval*

Exercises

1. Explain that for precise data the definition of this section yields the indicator function of the classical confidence set.

2. For interval data the result of the procedure in this section is an indicator function. Is there a difference for the generalized confidence set under the different combination-rules?

16 Nonparametric estimation

The most important nonparametric estimation of the probability distribution of a one-dimensional stochastic quantity X, based on a sample $X_1, \ldots, X_n$ is the *empirical distribution function* $\hat{F}_n(\cdot)$ defined by

$$\hat{F}_n(x) = \frac{1}{n} \sum_{i=1}^{n} I_{(-\infty, x]}(X_i) \quad \text{for all} \ x \in \mathbb{R}.$$

For non-precise observations $x_1^\star, \ldots, x_n^\star$ of X different generalizations are possible. One possibility is explained in section 9. Other generalizations are given in the following.

16.1 Smoothed empirical distribution function

The classical empirical distribution function is a step function. In estimating continuous distributions it would be good to have a continuous estimate.

For non-precise data $x_1^\star, \ldots, x_n^\star$ with characterizing functions $\xi_1(\cdot), \ldots, \xi_n(\cdot)$ in natural way a continuous estimate $F_n^\star(\cdot)$ of the distribution function is given by

$$F_n^\star(x) = \frac{1}{n} \sum_{i=1}^{n} \frac{\displaystyle\int_{-\infty}^{x} \xi_i(t)dt}{\displaystyle\int_{-\infty}^{\infty} \xi_i(t)dt} \quad \forall \ x \in \mathbb{R}.$$

This smoothed e.d.f. is defined only for all observations being non-precise.

In figure 16.1 a non-precise sample and the corresponding estimate $F_n^\star(\cdot)$ is depicted.

16.2   Interval-valued empirical distribution function

For interval data in section 9.1 a generalized empirical distribution function $\hat{F}_n(\cdot \mid x_1^\star, \ldots, x_n^\star)$ is explained. This interval valued function is bounded by two functions $\underline{F}_n(\cdot)$ and $\overline{F}_n(\cdot)$ with

$$\underline{F}_n(x) \leq \overline{F}_n(x) \qquad \forall \; x \in \mathbb{R}.$$

The generalized values of $\hat{F}_n(\cdot \mid x_1^\star, \ldots, x_n^\star)$ for argument x are intervals

$$F_n(x \mid x_1^\star, \ldots, x_n^\star) = \left[\underline{F}_n(x), \; \overline{F}_n(x) \right].$$

For an example see figure 9.1 in section 9.

For general non-precise observations $x_1^\star, \ldots, x_n^\star$ with characterizing functions $\xi_1(\cdot), \ldots, \xi_n(\cdot)$ the two functions $\underline{F}_n(\cdot)$ and $\overline{F}_n(\cdot)$ are defined in the following way.

In case of pairwise disjoint supports $supp(\xi_i(\cdot))$ the characterizing functions can be ordered and the corresponding characterizing functions are denoted by $\xi_{(i)}(\cdot)$, $i = 1(1)n$.

In the interval $supp\left(\xi_{(i)}(\cdot)\right)$ the functions $\underline{F}_n(.)$ and $\overline{F}_n(.)$ are given by

$$\overline{F}_n(x) \;=\; \begin{cases} \left(i - 1 + \xi_{(i)}(x)\right)/n & \forall \; x : \; \xi_{(i)}(x) \uparrow \\[2mm] i/n & \forall \; x : \; \xi_{(i)}(x) = 1 \;\vee \\ & \qquad\;\; \xi_{(i)}(x) \downarrow \end{cases}$$

$$\underline{F}_n(x) \;=\; \begin{cases} (i-1)/n & \forall \; x : \; \xi_{(i)}(x) \uparrow \;\; \vee \\ & \qquad\;\; \xi_{(i)}(x) = 1 \\[2mm] \left(i - \xi_{(i)}(x)\right)/n & \forall \; x : \; \xi_{(i)}(x) \downarrow \end{cases}$$

In figure 16.2 the section of these two functions in the interval $supp\left(\xi_{(i)}(\cdot)\right)$ is depicted and for a sample of four non-precise observations the whole interval valued exmpirical distribution function $\hat{F}_4(\cdot \mid x_1^\star, \ldots, x_4^\star)$ is drawn.

$\xi_i(x)$

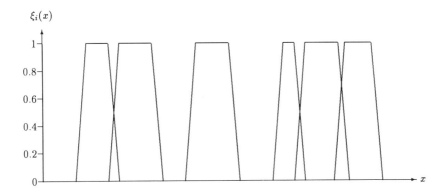

$F_n^\star(x)$

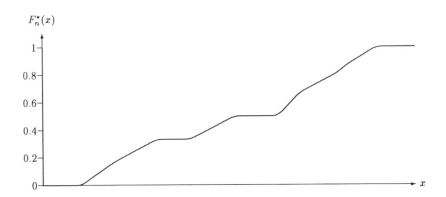

Figure 16.1: *Non-precise observations and smoothed empirical distribution function* $F_n^\star(\cdot)$

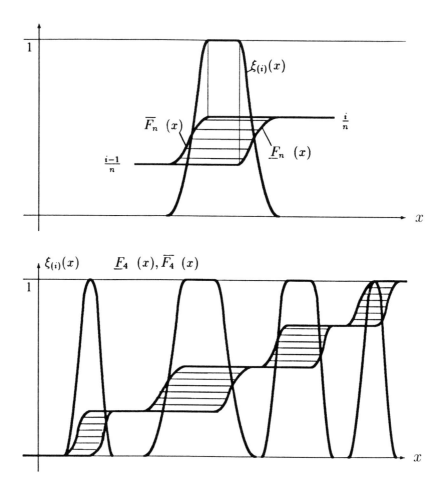

Figure 16.2: *Interval valued empirical distribution function*
$$\hat{F}_n(\cdot \mid x_1^\star, \ldots, x_n^\star)$$

Remark 16.1: For non-precise data with intersecting supports of the characterizing functions a superposition of the above-defined functions $\underline{F}_n(\cdot)$ and $\overline{F}_n(\cdot)$ is constructed. An example is given in figure 16.3.

16.3 Application of the propagation of imprecision

For a sample of n non-precise observations of a one-dimensional stochastic quantity X let $\xi(\cdot,\ldots,\cdot)$ be the characterizing function of the non-precise combined sample element.

The classical empirical distribution function $\hat{F}_n(\cdot)$ whose values are given by

$$\hat{F}_n(x) = \frac{1}{n}\sum_{i=1}^{n}I_{(-\infty,x]}(x_i)$$

cannot be used directly to construct a generalized empirical distribution function.

A generalization is possible using the inverted empirical distribution function

$$\hat{F}^{-1}(k,x_1,\cdots,x_n) := \inf\left\{z \in \mathbb{R} : \hat{F}_n(z) = \frac{k}{n}\right\}.$$

This function is continuous in the observations $x_1,\cdots,x_n$. Therefore we obtain for non-precise data $x_1^*,\cdots,x_n^*$ a function

$$\left(\hat{F}^{-1}\right)^*(k,x_1,\cdots,x_n)$$

with non-precise values.

The generalized inverted non-precise empirical distribution function is defined by its non-precise values $\left(\hat{F}^{-1}\right)^*(k)$ whose characterizing function $\varphi_{(\hat{F}^{-1})^*(k)}(\cdot)$ is given by its values

$$\varphi_{(\hat{F}^{-1})^*(k)}(z) = \sup_{\underline{x}\in M^n}\left\{\xi(\underline{x}) : \hat{F}^{-1}(k,\underline{x}) = z\right\} \qquad \text{for } k = 1(1)n$$

$$\forall \ z \in \mathbb{R}.$$

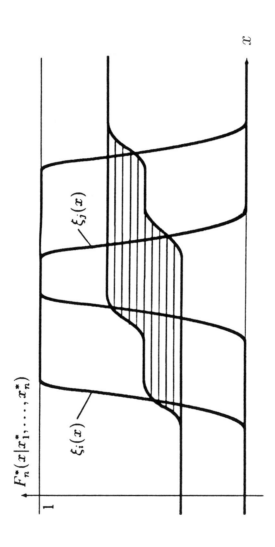

Figure 16.3: *Interval valued empirical distribution function for intersecting non-precise data*

The α-level curves of the so-defined generalized empirical distribution function are given by

$$\left(\hat{F}^{-1}\right)_\alpha^U (k) = \max_{\underline{x} \in B_\alpha(\underline{x}^*)} \hat{F}^{-1}(k, \underline{x})$$
$$\left(\hat{F}^{-1}\right)_\alpha^L (k) = \min_{\underline{x} \in B_\alpha(\underline{x}^*)} \hat{F}^{-1}(k, \underline{x})$$
$$\text{for} \quad k = 1(1)n.$$

Remark 16.2: If the minimum combination-rule is used, the *generalized corresponding empirical distribution function* can be represented by its upper and lower α-level curves $\hat{F}_\alpha^U(\cdot)$ and $\hat{F}_\alpha^L(\cdot)$ which are given by

$$\hat{F}_\alpha^U(z) = \frac{1}{n} \sum_{i=1}^n I_{(-\infty, z]}\left(\underline{B}_\alpha(x_i^*)\right)$$

$$\hat{F}_\alpha^L(z) = \frac{1}{n} \sum_{i=1}^n I_{(-\infty, z]}\left(\overline{B}_\alpha(x_i^*)\right)$$

where $B_\alpha(x_i^*) = \left[\underline{B}_\alpha(x_i^*), \overline{B}_\alpha(x_i^*)\right]$ are the α-cuts of the observations x_i^*.

An example is given in figure 16.4

16.4 Graphical generalization of the e.d.f.

For a sample $x_1^*, \cdots, x_n^*$ of non-precise observations of a one-dimensional stochastic quantity X a graphical generalization of the empirical distribution function can be obtained in the following way. Let $\xi_1(\cdot), \cdots, \xi_n(\cdot)$ be the characterizing functions corresponding to the non-precise observations $x_1^*, \cdots, x_n^*$. Ordering the functions $\xi_1(\cdot), \cdots, \xi_n(\cdot)$ by the left boundaries of the supports $supp(\xi(\cdot))$ and denoting the ordered set by $\xi_{(1)}(\cdot), \cdots, \xi_{(n)}(\cdot)$ a generalization of the classical empirical distribution function $\hat{F}_n(\cdot)$ can be obtained as given in figure 16.5.

Remark 16.3: In case all characterizing functions $\xi_i(\cdot)$ have the same shape, i.e., they can be considered as translations from each other, the α-level curves of the graphical generalization $F_n^*(\cdot \mid \xi_1(\cdot), \cdots, \xi_n(\cdot))$ have the form of classical empirical distribution functions.

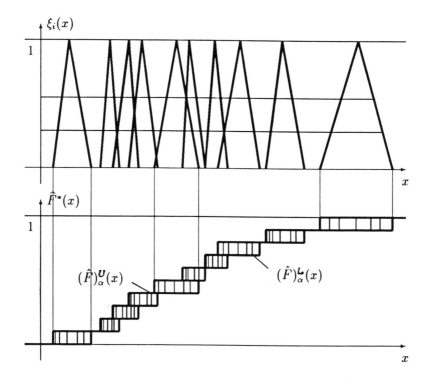

Figure 16.4: α-Level curves of the generalized e.d.f. using the
propagation of imprecision

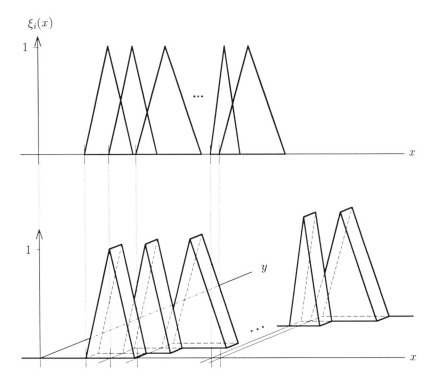

Figure 16.5: *Graphical generalization of the e.d.f.*

16.5 Empirical correlation coefficient for non-precise observations

Let $(x_i, y_i)^\star, i = 1, \ldots, n$ be n non-precise two-dimensional observations with corresponding characterizing functions

$$\xi_i(x, y) \text{ with } (x, y) \in \mathbb{R}^2.$$

The classical empirical correlation coefficient $r_{x,y}$ for precise observations (x_i, y_i), $i = 1, \ldots, n$,

$$r_{x,y} = \frac{\sum_{i=1}^{n}(x_i - \overline{x}_n)(y_i - \overline{y}_n)}{\sqrt{\left[\sum_{i=1}^{n}(x_i - \overline{x}_n)^2\right] \cdot \left[\sum_{i=1}^{n}(y_i - \overline{y}_n)^2\right]}}$$

can be generalized to the situation of non-precise data in the following way:

The n non-precise observations have to be combined to a non-precise element of the sample space $\mathbb{R}^{2n}$. The characterizing function of this element is denoted by

$$\varphi\left(\underline{(x, y)}\right) = \varphi(x_1, y_1, x_2, y_2, \ldots, x_n, y_n)$$

and can be obtained by a suitable combination of the characterizing functions $\xi_i(x, y)$.

Possible combinations are

$$\varphi\left(\underline{(x, y)}\right) := \min_{i=1(1)n} \xi_i(x_i, y_i)$$

and

$$\varphi\left(\underline{(x, y)}\right) := \prod_{i=1}^{n} \xi_i(x_i, y_i).$$

The generalized empirical correlation coefficient is the non-precise number $r^\star$, called *fuzzy correlation coefficient*, based on a non-precise sample $(x_i, y_i)^\star$, $i = 1(1)n$, defined by its characterizing function $\psi_{r^\star}(.)$ with values given by

$$\psi_{r^\star}(r) = \sup\left\{\varphi\left(\underline{(x, y)}\right) : r_{x,y} = r\right\}.$$

Remark 16.4: The non-precise observations can be also in the form $(x_i^\star, y_i^\star)$, $i = 1, \ldots, n$ with corresponding characterizing functions $\xi_{x_i^\star}(\cdot)$ and $\xi_{y_i^\star}(\cdot), i = 1, \ldots, n$. Moreover there can be also precise observations $x_i \in \mathrm{I\!R}$ and $y_i \in \mathrm{I\!R}$.

In figure 16.6 a non-precise sample and the corresponding fuzzy correlation coefficient is drawn.

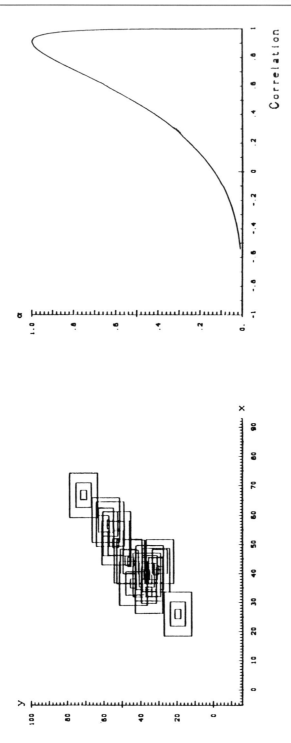

Figure 16.6: *Non-precise two-dimensional observations and characterizing function of the fuzzy correlation coefficient*

Exercises

1. Explain the difference of the smoothed empirical distribution function from section 16.1 and the cumulative sum from section 8.

2. Show the relationship between the generalized empirical distribution function for interval data from section 9 and the interval valued empirical distribution function from section 16.2.

3. What shape has the characterizing function of the empirical correlation coefficient in case of interval data?

17 Statistical tests and non-precise data

In classical significance testing based on precise observations $x_1, \ldots, x_n$ of a stochastic quantity $X \sim F_\theta$, $\theta \in \Theta$, and observation space M_X the decision is depending on the value of a *test statistic* $T = \tau(X_1, \ldots, X_n)$ for sample $X_1, \ldots, X_n$ of X.

For non-precise observations $x_1^\star, \ldots, x_n^\star$ with non-precise combined sample element $\underline{x}^\star$ and corresponding characterizing function $\xi(\cdot, \ldots, \cdot)$ the value of the test statistic becomes non-precise, modelled by definition 5.2. The characterizing function $\psi(.)$ of this non-precise value $t^\star$ of the test statistic $T = \tau(x_1, \ldots, x_n)$ is given by its values

$$\psi(t) = \sup_{\underline{x} \in M_X^n} \left\{ \xi(x_1, \ldots, x_n) : \tau(x_1, \ldots, x_n) = t \right\}. \qquad (17.1)$$

Example 17.1: For a sample $X_1, \ldots, X_n$ of a normally distributed stochastic quantity $X \sim N(\mu, \sigma^2)$ a test statistic for the hypothesis $\mathcal{H}_0 : \mu = \mu_0$ is

$$T = \tau(X_1, \ldots, X_n) = \frac{\overline{X}_n - \mu_0}{S_n / \sqrt{n}} \qquad (17.2)$$

and the acceptance region A for T under probability δ for an error of the first type is

$$A = \left\{ t \in \mathbb{R} : |t| = \left| \frac{\overline{x}_n - \mu_0}{s_n / \sqrt{n}} \right| \leq t_{n-1;1-\frac{\delta}{2}} \right\}$$

where $t_{n-1;1-\frac{\delta}{2}}$ is the $(1 - \frac{\delta}{2})$-fractile of the t-distribution with $n - 1$ degrees of freedom.

For non-precise data $x_1^\star, \ldots, x_n^\star$ the value of the test statistic becomes a non-precise number $t^\star$ using equations (17.1) and (17.2).

In figure 17.1 two possible characterizing functions of non-precise values $t^\star$ are depicted.

If the support of $t^\star$ is a subset of A or a subset of the complement A^c of A then a decision on acceptance or rejection of the hypothesis is possible as for exact observations.

In case that the support of $t^\star$ has non-empty intersection with both A and A^c an immediate decision is not possible. In this case similar to sequential tests more observations are necessary.

Remark 17.1: There are proposals for generalizations of statistical tests for non-precise observations. These papers are not published yet but are available from the author.

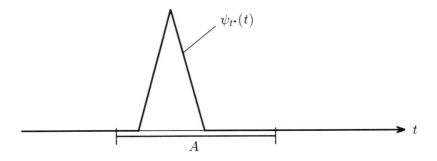

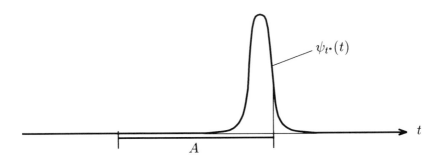

Figure 17.1: *Non-precise values $t^\star$ of the test statistic*

Exercises

In the following exercises use ten equally spaced α-cuts for the calculations.

1. Calculate the non-precise value of the test statistic from example 17.1 for the following non-precise sample using the minimum combination-rule.

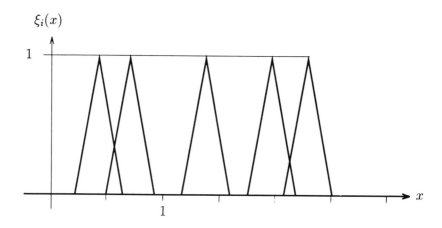

Draw the result together with the critical values for the test.

2. Consider the problem of exercise 1 but use the product combination-rule to obtain the characterizing functions of the non-precise combined sample element.

Chapter V

Bayesian inference for non-precise data

18 Bayes' theorem for non-precise data

For continuous stochastic model $X \sim f(\cdot \mid \theta)$, $\theta \in \Theta$ and continuous parameter space Θ and *a priori* density $\pi(\cdot)$ of the parameter and observation space M_X of X, Bayes' theorem for precise data $x_1, \ldots, x_n$ is

$$\pi(\theta \mid x_1, \ldots, x_n) = \frac{\pi(\theta)l(\theta; x_1, \ldots, x_n)}{\int_\Theta \pi(\theta)l(\theta; x_1, \ldots, x_n)d\theta} \qquad \forall \ \theta \in \Theta,$$

where $l(; x_1, \ldots, x_n)$ is the likelihood function. In the most simple situation of complete data the likelihood function is given by

$$l(\theta; x_1, \ldots, x_n) = \prod_{i=1}^{n} f(x_i \mid \theta) \qquad \forall \ \theta \in \Theta.$$

Remark 18.1: Using the abbreviation $\underline{x} = (x_1, \ldots, x_n)$ Bayes' theorem can be stated in the form

$$\pi(\theta \mid \underline{x}) \propto \pi(\theta) \cdot l(\theta; \underline{x}) \qquad \forall \ \theta \in \Theta,$$

where $\propto$ stands for "proportional" since the right hand of the formula is a non-normalized function which is — after normalization — a density on the parameter space Θ.

For non-precise data $D^\star = (x_1^\star, \ldots, x_n^\star)$ with corresponding characterizing functions $\xi_1(\cdot), \ldots, \xi_n(\cdot)$ the non-precise combined sample element $\underline{x}^\star$ with characterizing function $\xi(\cdot, \ldots, \cdot)$,

$$\xi : M_X^n \to [0, 1],$$

is the basis for a generalization of Bayes' theorem to the situation of non-precise data in the following way.

For all $\underline{x} \in supp(\xi(\cdot, \ldots, \cdot))$ the value $\pi(\theta|\underline{x})$ of the *a posteriori* density $\pi(\cdot \mid \underline{x})$ is calculated using Bayes' theorem for precise data.

To every θ by variation of $\underline{x}$ in the support of the non-precise combined sample element $\underline{x}^\star$ a family

$$\left(\pi(\theta|\underline{x}); \ \underline{x} \in supp(\underline{x}^\star)\right)$$

of values is obtained and the characterizing function $\psi_\theta(\cdot)$ of this non-precise value is obtained via the characterizing function $\xi(\cdot, \ldots, \cdot)$ of the non-precise combined sample element by its values

$$\psi_\theta(y) := \ \sup \ \{\xi(\underline{x}): \ \pi(\theta \mid \underline{x}) = y\} \tag{18.1}$$

where the supremum has to be taken over the sample space M_X^n.

The family $\left(\psi_\theta(\cdot), \ \theta \in \Theta\right)$ of non-precise values of the *a posteriori* distribution is describing the imprecision of the observations $x_i^\star$, $i = 1, \ldots, n$ and is called *non-precise* a posteriori *density* $\pi^\star(\cdot \mid D^\star)$, i.e.,

$$\pi^\star(\cdot \mid D^\star) := \left(\psi_\theta(\cdot); \ \theta \in \Theta\right).$$

A graphical presentation of the non-precise *a posteriori* distribution is the drawing of so-called α-level curves. These α-*level curves* are the curves which connect the ends of the α-cuts of $\psi_\theta(\cdot)$ as functions of θ. An example of the presentation of a non-precise *a posteriori* density is given in figure 18.1.

Example 18.1: Let X be a stochastic quantity having exponential distribution with density

$$f(x \mid \theta) = \frac{1}{\theta} \ e^{-x/\theta} I_{(0,\infty)}(x) \text{ with } \theta \in (0, \infty) = \Theta,$$

and the *a priori* density $\pi(\cdot)$ a gamma-density

$$\pi(\theta) = \frac{\theta^{\alpha-1}e^{-\theta/\beta}}{\Gamma(\alpha)\beta^\alpha} I_{(0,\infty)}(\theta) \quad \text{with } \alpha > 0, \ \beta > 0,$$

$\Gamma(\cdot)$ denoting the gamma-function defined by

$$\Gamma(\alpha) = \int_0^\infty t^\alpha e^{-t}dt.$$

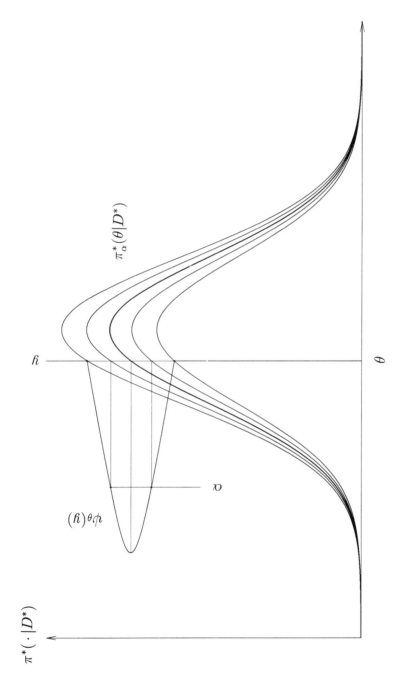

Figure 18.1: α-Level curves of a non-precise a posteriori distribution

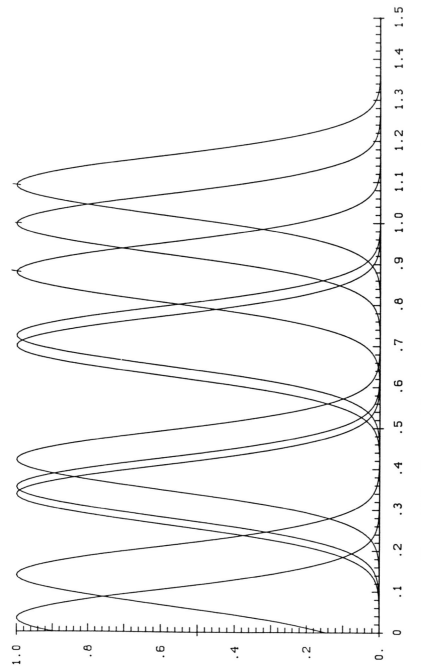

Figure 18.2: *Characterizing functions of ten non-precise observations*

Assume ten non-precise observations $x_1^\star, \ldots, x_n^\star$ of X are given by their characterizing functions in figure 18.2.

In order to obtain the non-precise *a posteriori* density $\pi^\star(\cdot \mid x_1^\star, \ldots, x_n^\star)$ the minimum combination-rule is used to obtain the characterizing function $\xi(\cdot, \ldots, \cdot)$ of the non-precise combined sample element $\underline{x}^\star$, i.e.,

$$\xi(x_1, \ldots, x_n) = \min_{i=1(1)n} \xi_i(x_i) \qquad \forall \; (x_1, \ldots, x_n) \in \mathbb{R}^n.$$

Using this, the characterizing functions $\psi_\theta(\cdot)$ of the non-precise values of the *a posteriori* density are obtained using equation (18.1) for every $\theta \in \Theta = (0, \infty)$.

In figure 18.3 some α-level curves of the non-precise *a posteriori* density $\pi^\star(\cdot \mid x_1^\star, \ldots, x_n^\star)$ are depicted.

Remark 18.2: The non-precise *a posteriori* density can be used for estimations and decisions. This will be explained in the following sections.

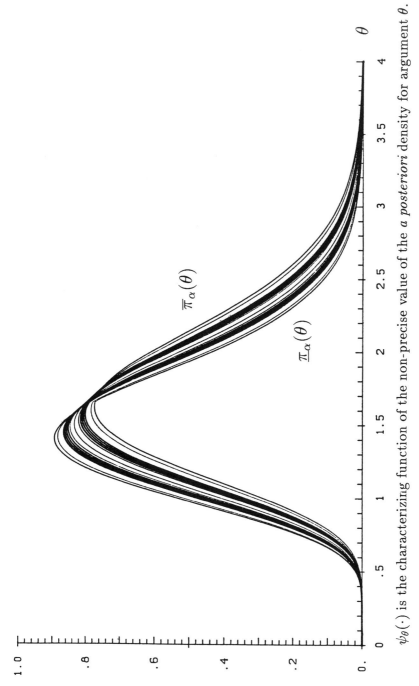

$\psi_\theta(\cdot)$ is the characterizing function of the non-precise value of the *a posteriori* density for argument θ.

Figure 18.3: α-*Level curves of the non-precise a posteriori density*

Exercises

1. Calculate the non-precise *a posteriori* density for the following problem. Let X be uniformly distributed on $(0, \theta)$ and the *a priori* distribution of $\tilde{\theta}$ also a uniform distribution with density $\frac{1}{b-a} I_{(a,b)}(\theta)$. For non-precise data with characterizing functions given below, calculate the α-level curves of the non-precise *a posteriori* density $\pi^\star(\theta \mid D^\star)$ for $\alpha = 0.1(0.1)1$.

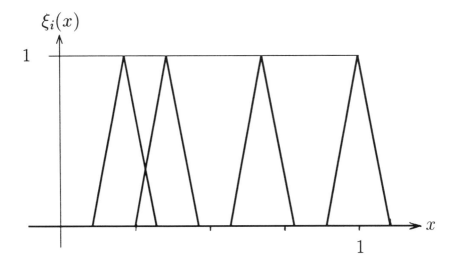

2. Let X be normal distributed with known variance σ^2 and unknown mean θ. For normal *a priori* distribution with hyperparameter vector (μ, τ) and non-precise sample consider the necessary calculations to obtain the non-precise *a posteriori* density of $\tilde{\theta}$.

19   Bayesian confidence regions based on non-precise data

In this section a stochastic model $X \sim f(\cdot | \theta)$, $\theta \in \Theta$ is used and for the parameter θ an *a priori* distribution $\pi(\cdot)$ is supposed to be given. The stochastic quantity describing the uncertain parameter θ is denoted by $\tilde{\theta}$.

Generalizing the Bayesian concept of confidence regions for non-precise data and non-precise *a posteriori* distributions non-precise Bayesian confidence regions can be constructed.

19.1   Generalized Bayesian confidence regions

For precise data $D = \underline{x} = (x_1, \ldots, x_n)$ and exact *a posteriori* density $\pi(\cdot \mid \underline{x})$ a *Bayesian confidence region* Θ_0 for θ with confidence level $1 - \delta$ is defined by

$$Pr_{\pi(\cdot | \underline{x})} \left\{ \tilde{\theta} \in \Theta_0 \right\} = \int_{\Theta_0} \pi(\theta \mid \underline{x}) \, d\theta \; = \; 1 - \delta. \qquad (19.1)$$

In case of non-precise data $D^\star = (x_1^\star, \ldots, x_n^\star)$ generalized confidence regions for θ can be constructed using the non-precise combined sample element $\underline{x}^\star$.

Definition 19.1: Let $\xi(\cdot, \ldots, \cdot)$ be the characterizing function of the non-precise combined sample element $\underline{x}^\star$. Then for every $\underline{x} \in supp(\xi(\cdot))$ a Bayesian confidence region $\Theta_{\underline{x}}$ is calculated using equation (19.1). The *generalized Bayesian confidence region* for θ

with confidence level $1 - \delta$ is the *fuzzy subset* $\Theta^\star$ of Θ whose characterizing function $\psi(\cdot)$ is given by its values

$$\psi(\theta) := \sup_{\underline{x} \in M_X^n} \left\{ \xi(\underline{x}) : \theta \in \Theta_{\underline{x}} \right\} \qquad \forall \ \theta \in \Theta.$$

Remark 19.1: The generalized Bayesian confidence regions are reasonable generalizations by the following inequality:

$$I_{\displaystyle\bigcup_{\underline{x}:\ \xi(\underline{x})=1} \Theta_{\underline{x}}}(\theta) \leq \psi(\theta) \qquad \forall \ \theta \in \Theta.$$

This means that $\psi(\cdot)$ dominates the indicator functions of all classical Bayesian confidence regions $\Theta_{\underline{x}}$ with $\xi(\underline{x}) = 1$.

19.2   Generalized HPD-regions

Highest *a posteriori* density regions, abbreviated by HPD-regions, for the parameter θ of a stochastic model $X \sim f(\cdot \mid \theta)$, $\theta \in \Theta$ with continuous parameter θ and precise data $D = \underline{x} = (x_1, \ldots, x_n)$ are defined using the *a posteriori* density $\pi(\cdot \mid \underline{x})$.

The *HPD-region* for θ with confidence level $1 - \delta$ based on precise data $\underline{x}$ and *a posteriori* density $\pi(\cdot \mid \underline{x})$ is the subset Θ_0 of Θ, for which holds

$$(1) \qquad \int_{\Theta_0} \pi(\theta \mid \underline{x}) \, d\theta \ = \ 1 - \delta$$

and

$$(2) \qquad \pi(\theta \mid \underline{x}) \geq C \quad \text{for all } \theta \in \Theta_0,$$

where C is the largest possible constant for which (1) and (2) hold.

Remark 19.2: HPD-regions are Bayesian confidence regions which are as small as possible and mostly unique. The generalization of HPD-regions to the situation of non-precise *a posteriori* distributions is possible in the following way.

     For non-precise data $D^\star = (x_1^\star, \ldots, x_n^\star)$ let $\xi(\cdot, \ldots, \cdot)$ be the characterizing function of the non-precise combined sample element $\underline{x}^\star$. Then for every precise vector $\underline{x}$ in $supp(\xi(\cdot, \ldots, \cdot))$ the *a posteriori* density $\pi(\cdot \mid \underline{x})$ can be calculated.

Definition 19.2: In order to construct a generalized HPD-region which is a fuzzy subset of the parameter space Θ, for all $\theta \in \Theta$ and $\underline{x} \in supp(\xi(\cdot))$, a classical subset $B(\theta, \underline{x})$ of Θ is defined by

$$B(\theta, \underline{x}) := \{\theta' \in \Theta : \pi(\theta' \mid \underline{x}) \geq \pi(\theta \mid \underline{x})\}. \qquad (19.2)$$

The *generalized HPD-region* for θ with confidence level $1 - \delta$ is the *fuzzy subset* of Θ defined by its characterizing function $\varphi(\cdot)$ given by its values

$$\varphi(\theta) = \sup \left\{ \xi(\underline{x}) : \int_{B(\theta, \underline{x})} \pi(\theta' \mid \underline{x}) \, d\theta' \leq 1 - \delta \right\} \qquad (19.3)$$

where the supremum has to be taken over the set $supp(\xi(\cdot))$.

Remark 19.3: The construction of generalized HPD-regions applied to precise data, describing the precise data by its one-point indicator functions, yields as a result the indicator function of the classical HPD-region for precise data.

Example 19.1: For stochastic quantity X with exponential distribution having density

$$f(x \mid \theta) = \frac{1}{\theta} e^{-x/\theta} I_{(0,\infty)}(x)$$

and ten non-precise observations $x_1^\star, \ldots, x_{10}^\star$ whose characterizing functions are presented in figure 19.1, the minimum combination-rule is used to obtain the characterizing function of the non-precise combined sample element.

Application of the definition of the characterizing function $\varphi(\cdot)$ from equation (19.3) yields the fuzzy HPD-region with characterizing function shown in figure 19.2.

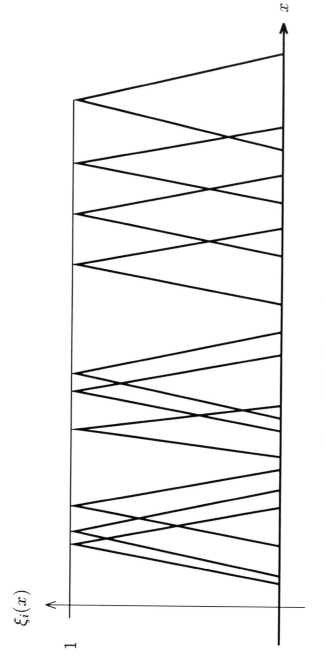

Figure 19.1: *Non-precise observations*

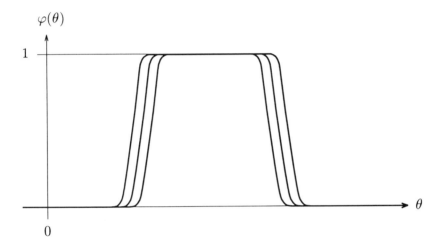

Characterizing functions for different confidence levels $1 - \delta$

Figure 19.2: *Fuzzy HPD-regions for θ*

Exercises

1. Assume that a stochastic quantity has normal distribution $N(\theta, \sigma_0^2)$. For non-precise sample given in the following figure, use the minimum combination-rule to construct the non-precise combined sample element $\underline{x}^*$.

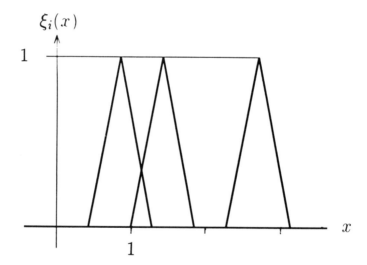

Let σ_0^2 be known and the *a priori* distribution $\pi(\theta)$ be a normal distribution with hyperparameters η, τ^2, i.e., $\tilde{\theta} \sim N(\eta, \tau^2)$. Use the fact that the *a posteriori* distribution $\pi(\theta \mid \underline{x})$ is a normal distribution for all precise data $\underline{x} = (x_1, \ldots, x_n)$ to construct generalized Bayesian confidence regions for θ. Hint: Show first that $\pi(\theta \mid x_1, \ldots, x_n)$ is

$$
N\left(\frac{\eta \sigma_0^2}{n\tau^2 + \sigma_0^2} + \frac{\tau^2 \sum_{i=1}^{n} x_i}{n\tau^2 + \sigma_0^2}, \frac{\tau^2 \sigma_0^2}{n\tau^2 + \sigma_0^2} \right).
$$

2. Take the data from exercise 1 but use the product combination-rule to obtain the non-precise combined sample element.
   Then work out the exercise as in exercise 1 and compare the
   results.

20   Non-precise predictive distributions

Information on future values of stochastic quantities X with observation space M_X and parametric stochastic model $f(\cdot \mid \theta)$, $\theta \in \Theta$ is provided by the predictive density.

In case of precise data $\underline{x} = (x_1, \ldots, x_n)$ and corresponding *a posteriori* density $\pi(\cdot \mid \underline{x})$ for the parameter $\tilde{\theta}$ the predictive density $g(\cdot \mid \underline{x})$ for X conditional on data $\underline{x}$ is the conditional density of X, i.e.,

$$g(x \mid \underline{x}) = \int_\Theta f(x \mid \theta)\, \pi(\theta \mid \underline{x})\, d\theta \quad \text{for all} \quad x \in M_X. \quad (20.1)$$

For non-precise data $D^\star = (x_1^\star, \ldots, x_n^\star)$ the non-precise combined sample element $\underline{x}^\star$ with characterizing function

$$\xi(\cdot, \ldots, \cdot)$$

is used for the generalization of the concept of predictive densities. This generalization is defined by a family of non-precise values for the predictive density.

Definition 20.1: For fixed $x \in M_X$ the vector $\underline{x}$ is varying in $supp(\xi(\cdot, \ldots, \cdot))$ and the characterizing function $\psi_x(\cdot)$ of the *non-precise value of the predictive density* is given by its values

$$\psi_x(y) = \sup \left\{ \xi(\underline{x}) : \underline{x} \in M_X^n, g(x \mid \underline{x}) = y \right\} \quad \forall \ x \in M_X,$$

where $g(\cdot \mid \underline{x})$ is the value of the classical predictive density based on precise data $\underline{x}$ and the supremum is to be taken over the set $supp(\xi(\cdot, \ldots, \cdot))$.

The family $(\psi_x(\cdot),\ x \in M_X)$ of non-precise values of the predictive density is called *non-precise predictive density*

$$g^\star(\cdot \mid D^\star) = (\psi_x(\cdot);\ x \in M_X).$$

Remark 20.1: A graphical representation of non-precise predictive densities can be given using α-level curves which are described in section 18.

Example 20.1: In continuation of example 18.1 the non-precise predictive density $g^\star(\cdot \mid D^\star)$ can be calculated. A representation of the result using α-level curves is given in figure 20.1.

Remark 20.2: For precise data $\underline{x} = (x_1,\ldots,x_n)$ with $\xi_i(\cdot) = I_{\{x_i\}}(\cdot)$ the resulting characterizing functions are

$$\psi_x(\cdot) = I_{\{g(x\mid\underline{x})\}}(\cdot).$$

Therefore, the concept is a reasonable generalization of the classical predictive density.

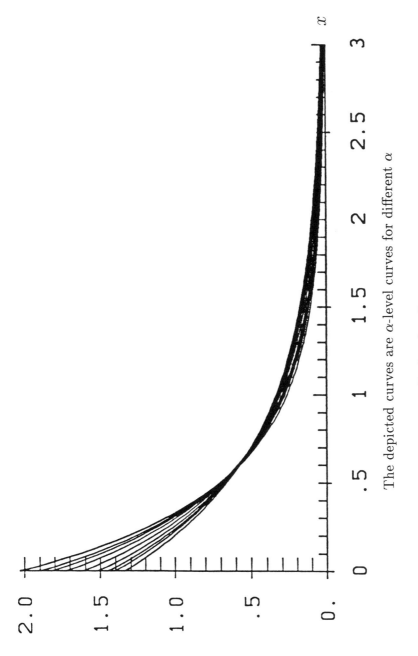

The depicted curves are α-level curves for different α

Figure 20.1: *Non-precise predictive density*

Exercises

1. Let X be exponentially distributed with density

$$f(x \mid \tau) = \frac{1}{\tau} e^{-x/\tau} I_{(0,\infty)}(x)$$

   with *a priori* density

$$\pi(\tau) = I_{(0,20)}(\tau) \qquad \forall \ \tau > 0.$$

   Take a non-precise sample of X of your choice and use the minimum combination-rule to obtain the non-precise predictive density for X.

2. Work out example 1 using the product combination-rule and compare the results.

21 Non-precise a priori distributions

Using precise *a priori* distributions for parameters θ in stochastic models

$$X \sim F_\theta, \quad \theta \in \Theta$$

is a topic of critical discussions. Allowing a more general formulation of *a priori* knowledge, general agreement could arise on reasonable use of *a priori* information on parameters.

Looking at the result in section 18 in natural way non-precise *a priori* distributions in form of non-precise densities $\pi^\star(\cdot)$ can be used. These non-precise densities are given by the family $\pi^\star(\theta)$, $\theta \in \Theta$ of fuzzy values of the density with characterizing functions $\varphi_\theta(\cdot)$, i.e.,

$$\pi^\star(\cdot) = \big(\pi^\star(\theta); \ \theta \in \Theta\big) \ \hat{=} \ \big(\varphi_\theta(\cdot); \ \theta \in \Theta\big).$$

This formulation is also necessary to describe the sequential information gaining process which is obtained by gathering additional data. Therefore, the modelling from section 18 has to be generalized in the following way.

Let $x_1^\star, \ldots, x_n^\star$ be n non-precise observations with corresponding characterizing functions $\xi_1(\cdot), \ldots, \xi_n(\cdot)$ and $\pi^\star(\cdot)$ a non-precise *a priori* distribution with non-precise values $\pi^\star(\theta)$ and corresponding characterizing functions $\varphi_\theta(\cdot)$. Then the imprecision of the *a priori* distribution has to be combined with the imprecision of the non-precise combined sample element $\underline{x}^\star$ with characterizing function $\xi(\cdot, \ldots, \cdot)$.

This combination and the generalization must yield a non-precise *a posteriori* distribution for $\tilde{\theta}$.

In case of non-precise *a priori* distributions formed by non-precise parameters of the *a priori* distribution the analysis is relatively simple. This will be explained in the following subsection.

21.1   Non-precise hyperparameters

In many situations the imprecision of an *a priori* distribution can be expressed by the imprecision of a parameter of the *a priori* distribution. This is especially valid for conjugate families of distributions.

Let $\pi(\cdot)$ be the *a priori* distribution for the parameter θ of the stochastic model

$$X \sim f(\cdot \mid \theta),\ \theta \in \Theta.$$

In this section it is assumed that $\pi(\cdot)$ is determined by a so-called *hyperparameter* λ, i.e.,

$$\pi(\cdot) \ = \ \pi(\cdot \mid \lambda), \quad \lambda \in \Lambda.$$

The imprecision of the *a priori* distribution can be modelled by an imprecision of λ, i.e., $\lambda^\star$ is a non-precise element of Λ with characterizing function $\chi(\cdot)$.

In order to describe the non-precise *a posteriori* density the *non-normalized a posteriori densities*

$$g_n(\theta \mid \lambda, \underline{x}) \ = \ \pi(\theta \mid \lambda) \cdot l(\theta; \underline{x})$$

for precise data $\underline{x} = (x_1, \ldots, x_n)$ are helpful.

The functions $g_n(\cdot \mid \lambda, \underline{x})$ are related to classical Bayes' theorem by

$$\pi(\theta \mid \lambda, \underline{x}) \ = \ \frac{g_n(\theta \mid \lambda, \underline{x})}{\displaystyle\int_\Theta g_n(\theta \mid \lambda, \underline{x})d\theta} \ .$$

For *non-precise hyperparameter* $\lambda^\star$ of the *a priori* distribution, the non-precise *a priori* distribution $\pi^\star(\cdot \mid \lambda^\star)$ is given by its α-level curves $\underline{\pi}_\alpha(\cdot)$ and $\overline{\pi}_\alpha(\cdot)$ in the following way:

$$\underline{\pi}_\alpha(\theta) = \min_{\lambda \in B_\alpha(\lambda^\star)} \pi(\theta \mid \lambda)$$

and

$$\overline{\pi}_\alpha(\theta) = \max_{\lambda \in B_\alpha(\lambda^*)} \pi(\theta \mid \lambda).$$

The α-level curves $\underline{g}_{n,\alpha}(\cdot)$ and $\overline{g}_{n,\alpha}(\cdot)$ of the *generalized non-normalized a posteriori density*

$$g_n^*(\cdot \mid \lambda^*, \underline{x})$$

for non-precise *a priori* hyperparameter λ^* and precise data

$$\underline{x} = (x_1, \ldots, x_n)$$

are given by their values

$$\underline{g}_{n,\alpha}(\theta) = l(\theta; \underline{x}) \cdot \min_{\lambda \in B_\alpha(\lambda^*)} \pi(\theta \mid \lambda)$$

and

$$\overline{g}_{n,\alpha}(\theta) = l(\theta; x) \cdot \max_{\lambda \in B_\alpha(\lambda^*)} \pi(\theta \mid \lambda).$$

With the above notations Bayes' theorem for non-precise hyperparameters can be written also in sequential form

$$\underline{g}_{n,\alpha}(\theta) = \underline{g}_{n-1,\alpha}(\theta) \cdot f(x_n \mid \theta)$$

and

$$\overline{g}_{n,\alpha}(\theta) = \overline{g}_{n-1,\alpha}(\theta) \cdot f(x_n \mid \theta)$$

with $\qquad \underline{g}_{0,\alpha}(\theta) = \underline{\pi}_\alpha(\theta) \quad$ and $\quad \overline{g}_{0,\alpha}(\theta) = \overline{\pi}_\alpha(\theta).$

Example 21.1: For a stochastic quantity X with exponential distribution having density

$$f(x \mid \theta) = \frac{1}{\theta} e^{-x/\theta} I_{(0,\infty)}(x), \quad \theta \in \Theta = (0, \infty)$$

a conjugate family of *a priori* distributions is the family of gamma distributions with hyperparameter(vector) $\underline{\lambda} = (\nu, \beta)$ and densities

$$\pi(\theta \mid \nu, \beta) = \frac{\beta^\nu}{\Gamma(\nu)} \theta^{\nu-1} e^{-\beta\theta} I_{(0,\infty)}(\theta).$$

For non-precise subparameter $\beta^\star$ the α-level curves $\underline{\pi}_\alpha(\cdot)$ and $\overline{\pi}_\alpha(\cdot)$ of the non-precise *a priori* density $\pi(\cdot \mid \nu, \beta^\star)$ are given, using the notations

$$\beta_{min}(\theta) = \begin{cases} \underline{B}_\alpha(\beta^\star) & \text{for } \theta \leq \nu/[b(\underline{B}_\alpha(\beta^\star), \overline{B}_\alpha(\beta^\star))] \\ \overline{B}_\alpha(\beta^\star) & \text{for } \theta > \nu/[b(\underline{B}_\alpha(\beta^\star), \overline{B}_\alpha(\beta^\star))] \end{cases} \quad (21.1)$$

with

$$b(\underline{B}_\alpha(\beta^\star), \overline{B}_\alpha(\beta^\star)) = [\overline{B}_\alpha(\beta^\star) - \underline{B}_\alpha(\beta^\star)] \ln \frac{\overline{B}_\alpha(\beta^\star)}{\underline{B}_\alpha(\beta^\star)}$$

and

$$\beta_{max}(\theta) = \begin{cases} \overline{B}_\alpha(\beta^\star) & \text{for } \theta < \nu/[\overline{B}_\alpha(\beta^\star)] \\ \nu/\theta & \text{for } \nu/[\overline{B}_\alpha(\beta^\star)] \leq \theta \leq \nu/[\underline{B}_\alpha(\beta^\star)] \\ \underline{B}_\alpha(\beta^\star) & \text{for } \theta > \nu/[\underline{B}_\alpha(\beta^\star)] \end{cases}$$

$$(21.2)$$

by the following equations:

$$\underline{\pi}_\alpha(\theta) = \frac{[\beta_{min}(\theta)]^\nu}{\Gamma(\nu)} \theta^{\nu-1} \exp[-\beta_{min}(\theta) \cdot \theta]$$

$$\overline{\pi}_\alpha(\theta) = \frac{[\beta_{max}(\theta)]^\nu}{\Gamma(\nu)} \theta^{\nu-1} \exp[-\beta_{max}(\theta) \cdot \theta].$$

This generalized *a priori* distribution is called *non-precise gamma distribution*. In figure 21.1 some α-level curves of a non-precise gamma distribution are depicted.

For precise data $\underline{x} = (x_1, \ldots, x_n)$ the α-level curves of the non-normalized non-precise *a posteriori* density $g_n(\cdot \mid \nu, \beta^\star, \underline{x})$ are given by the following equations:

$$\underline{g}_{n,\alpha}(\theta) = \frac{[\beta_{min}(\theta)]^\nu}{\Gamma(\nu)} \theta^{\nu+n-1} \exp\left(-\left[\beta_{min}(\theta) + \sum_{i=1}^n x_i\right] \cdot \theta\right)$$

$$\overline{g}_{n,\alpha}(\theta) = \frac{[\beta_{max}(\theta)]^\nu}{\Gamma(\nu)} \theta^{\nu+n-1} \exp\left(-\left[\beta_{max}(\theta) + \sum_{i=1}^n x_i\right] \cdot \theta\right).$$

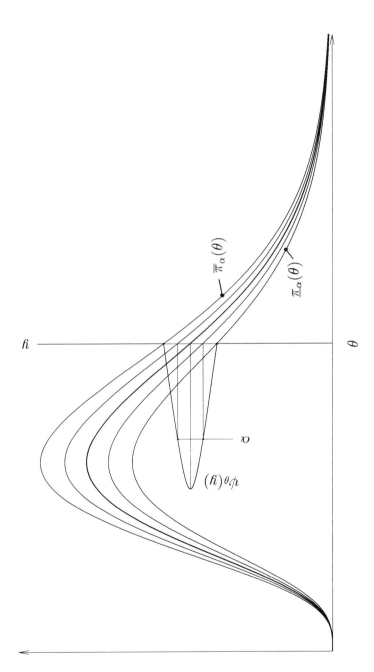

Figure 21.1: *Non-precise gamma density*

Exercises

1. Draw a diagram for a non-precise exponential distribution given by a non-precise value $\tau^\star$ of the parameter τ in the density

$$f(x \mid \tau) = \frac{1}{\tau} e^{-x/\tau} I_{(0,\infty)}(x).$$

Use α-level curves for the visualization of the non-precise density $f^\star(\cdot \mid \tau^\star)$.

2. Consider five non-precise observations with triangular characterizing functions from an exponentially distributed stochastic quantity. For non-precise parameter $\beta^\star$ of a non-precise gamma *a priori* distribution draw a diagram of four α-level curves of the non-normalized *a posteriori* density from subsection 21.1.

22   Bayes' theorem for non-precise a priori distribution and non-precise data

Let $X \sim f(\cdot \mid \theta)$, $\theta \in \Theta$ be a stochastic model and $\pi^\star(\cdot)$ a non-precise *a priori* distribution of the parameter θ. Moreover let $\pi^\star(\cdot)$ be given by a non-precise hyperparameter $\lambda^\star$ with characterizing function $\chi_{\lambda^\star}(\cdot)$, i.e.,

$$\pi^\star(\cdot) = \pi(\cdot \mid \lambda^\star) \quad \text{with} \quad \pi(\cdot \mid \lambda), \; \lambda \in \Lambda.$$

Bayes' theorem for non-precise observations $x_1^\star, \ldots, x_n^\star$ of X with non-precise combined sample element $\underline{x}^\star$ can be generalized in the following way.

Let $\xi_{\underline{x}^\star}(\cdot)$ be the characterizing function of $\underline{x}^\star$ and $\chi_{\lambda^\star}(\cdot)$ the characterizing function of the non-precise hyperparameter $\lambda^\star$. Then the imprecision of $\lambda^\star$ and that of $\underline{x}^\star$ have to be combined in order to obtain a non-precise element $(\lambda, \underline{x})^\star$ of the space $\Lambda \times M_X^n$.

For general combination-rules the inference can be complicated because the following sequential information processing property of Bayesian inference can be lost:

For precise data $D = (x_1, \ldots, x_n, x_{n+1}, \ldots, x_{n+k})$ and precise *a priori* distribution $\pi(\cdot)$ the calculation of $\pi(\cdot \mid D)$ by Bayes' theorem results in the same distribution when first calculating $\pi(\cdot \mid x_1, \ldots, x_n)$ and then using this distribution as a new *a priori* distribution $\pi_1(\cdot)$ and calculating the new *a posteriori* distribution $\pi_1(\cdot \mid x_{n+1}, \ldots, x_{n+k})$, i.e.,

$$\pi(\cdot \mid D) = \pi_1(\cdot \mid x_{n+1}, \ldots, x_{n+k}).$$

If the imprecision of $\lambda^\star$ and $\underline{x}^\star$ is combined with the minimum combination-rule, i.e., the characterizing function $\phi(\cdot, \cdot)$ of $(\lambda, \underline{x})^\star$ is given by

$$\phi(\lambda, \underline{x}) = \min \{\chi_{\lambda^\star}(\lambda), \, \xi_{\underline{x}^\star}(\underline{x})\},$$

then the calculation of the generalized non-normalized *a posteriori* density $g_n^\star(\cdot)$ after n observations can be obtained using the following relation for the α-cuts:

$$B_\alpha\big((\lambda, \underline{x})^\star\big) = B_\alpha(\lambda^\star) \times B_\alpha(\underline{x}^\star) \quad \forall \ \alpha \in (0, 1] \ .$$

Therefore, the α-level curves $\underline{g}_{n,\alpha}(\cdot)$ and $\overline{g}_{n,\alpha}(\cdot)$ of the generalized non-normalized *a posteriori* density of $\tilde{\theta}$ are given by

$$\underline{g}_{n,\alpha}(\theta) = \left(\min_{\lambda \in B_\alpha(\lambda^\star)} \pi(\theta \mid \lambda) \right) \cdot \left(\min_{\underline{x} \in B_\alpha(\underline{x}^\star)} l(\theta; \underline{x}) \right)$$

and

$$\overline{g}_{n,\alpha}(\theta) = \left(\max_{\lambda \in B_\alpha(\lambda^\star)} \pi(\theta \mid \lambda) \right) \cdot \left(\max_{\underline{x} \in B_\alpha(\underline{x}^\star)} l(\theta; \underline{x}) \right)$$

for all $\alpha \in (0, 1]$ and all $\theta \in \Theta$.

If the characterizing function of the non-precise combined sample element $\underline{x}^\star$ is also given by the minimum combination-rule then also in case of non-precise hyperparameter *and* non-precise data, sequential Bayesian inference is possible, i.e.,

$$\underline{g}_{n,\alpha}(\theta) = \underline{g}_{n-1,\alpha}(\theta) \cdot \min_{x \in B_\alpha(x_n^\star)} f(x \mid \theta)$$

$$\overline{g}_{n,\alpha}(\theta) = \overline{g}_{n-1,\alpha}(\theta) \cdot \max_{x \in B_\alpha(x_n^\star)} f(x \mid \theta),$$

with $\underline{g}_{0,\alpha}(\theta) = \underline{\pi}_\alpha(\theta)$ and $\overline{g}_{0,\alpha}(\theta) = \overline{\pi}_\alpha(\theta)$ being the α-level curves of the non-precise *a priori* density $\pi^\star(\cdot)$.

22.1   The α-level curves of non-precise a posteriori densities

Let $\pi(\cdot \mid \lambda)$ be the *a priori* density of $\tilde{\theta}$ with non-precise hyperparameter $\lambda^\star$ and

$$\big(B_\alpha(\lambda^\star); \alpha \in (0, 1]\big)$$

the α-cut presentation of $\lambda^\star$. Moreover let $\underline{x}^\star$ denote the non-precise combined sample element obtained from a sample $x_1^\star, \ldots, x_n^\star$ of non-precise observations. The α-cut representation of $\underline{x}^\star$ is

$$\left(B_\alpha(\underline{x}^\star); \ \alpha \in (0,1]\right).$$

Then the lower α-level curves $\underline{\pi}_\alpha^\star(\cdot \mid \underline{x}^\star)$ and the upper α-level curves $\overline{\pi}_\alpha^\star(\cdot \mid \underline{x}^\star)$ of the non-precise *a posteriori* density

$$\pi^\star(\cdot \mid \underline{x}^\star)$$

are given by the following equations:

$$\underline{\pi}_\alpha^\star(\theta \mid \underline{x}^\star) = \min_{(\lambda,\underline{x}) \in B_\alpha(\lambda^\star) \times B_\alpha(\underline{x}^\star)} \frac{g_n(\theta \mid \lambda, \underline{x})}{\displaystyle\int_\Theta g_n(\theta \mid \lambda, \underline{x}) d\theta}$$

$$\overline{\pi}_\alpha^\star(\theta \mid \underline{x}^\star) = \max_{(\lambda,\underline{x}) \in B_\alpha(\lambda^\star) \times B_\alpha(\underline{x}^\star)} \frac{g_n(\theta \mid \lambda, \underline{x})}{\displaystyle\int_\Theta g_n(\theta \mid \lambda, \underline{x}) d\theta}.$$

The functions $g_n(\theta \mid \lambda, \underline{x})$ are explained in section 21.1.

Remark 22.1: The minimization and maximization procedures can be complicated if the integration is not analytically possible. Essentially simpler is the analysis for conjugate families of distributions.

22.2 Conjugate families of distributions

In case of conjugate families of distributions the stochastic model

$$X \sim f(\cdot \mid \theta), \quad \theta \in \Theta$$

and the *a priori* family

$$\left(\pi(\cdot \mid \lambda), \ \lambda \in \Lambda\right) = \mathcal{P}$$

are related such that for every *a priori* distribution from $\mathcal{P}$ and all possible exact data $\underline{x} = (x_1, \ldots, x_n)$ the *a posteriori* distribution $\pi(\cdot \mid \lambda, \underline{x})$ belongs to $\mathcal{P}$, i.e.,

$$\pi(\cdot \mid \lambda, \underline{x}) = \pi(\cdot \mid \lambda_n)$$

with $\lambda_n \in \Lambda$ and $\lambda_n = s(\lambda, \underline{x})$ for a function $s(\cdot, \cdot)$.

Here λ denotes the *hyperparameter* of the *a priori* distribution.

Now let non-precise data $x_1^\star, \ldots, x_n^\star$ with non-precise combined sample element $\underline{x}^\star$ and α-cut representation

$$\left(B_\alpha(\underline{x}^\star); \quad \alpha \in (0,1] \right)$$

be given. Then the α-cut representation of the non-precise hyperparameter $\lambda_n^\star$ is given by

$$\left(B_\alpha(\lambda_n^\star); \quad \alpha \in (0,1] \right)$$

with

$$B_\alpha(\lambda_n^\star) = \bigcup_{\lambda \in B_\alpha(\lambda_0^\star)} \bigcup_{\underline{x} \in B_\alpha(\underline{x}^\star)} s(\lambda, \underline{x}) , \qquad (22.1)$$

where $\lambda_0^\star$ denotes the non-precise hyperparameter of the *a priori* distribution.

If the function $s(\cdot, \cdot)$ is continuous in both variables and $\lambda \in \mathbb{R}$ then the formula simplifies to an interval:

$$B_\alpha(\lambda_n^\star) = \left[\min_{\lambda \in B_\alpha(\lambda_0^\star)} \min_{\underline{x} \in B_\alpha(\underline{x}^\star)} s(\lambda, \underline{x}), \ \max_{\lambda \in B_\alpha(\lambda_0^\star)} \max_{\underline{x} \in B_\alpha(\underline{x}^\star)} s(\lambda, \underline{x}) \right].$$

Using the above notations the following theorem holds.

Theorem 22.1: Let $\lambda_n^\star$ be the non-precise hyperparameter of the *a posteriori* density whose α-cuts are given by equation (22.1), then the α-level curves of

$$\pi^\star \left(\cdot \mid (\underline{x}, \lambda)^\star \right)$$

from section 22.1 are given by

$$\underline{\pi}_\alpha^\star(\theta \mid \underline{x}^\star) = \min_{\lambda_n \in B_\alpha(\lambda_n^\star)} \pi(\theta \mid \lambda_n)$$

and

$$\overline{\pi}_\alpha^\star(\theta \mid \underline{x}^\star) = \max_{\lambda_n \in B_\alpha(\lambda_n^\star)} \pi(\theta \mid \lambda_n)$$

for all $\theta \in \Theta$ and all $\alpha \in (0,1]$.

Proof: From the corresponding equation in section 22.1 we obtain for the lower α-level curve of the *a posteriori* density:

$$
\begin{aligned}
\underline{\pi}_\alpha(\theta \mid \underline{x}^\star) &= \min_{(\lambda_0,\underline{x}) \in B_\alpha(\lambda_0^\star) \times B_\alpha(\underline{x}^\star)} \frac{g_n(\theta \mid \lambda_0, \underline{x})}{\displaystyle\int_\Theta g_n(\theta \mid \lambda_0, \underline{x}) d\theta} \\[2mm]
&= \min_{(\lambda_0,\underline{x}) \in B_\alpha(\lambda_0^\star) \times B_\alpha(\underline{x}^\star)} \pi(\theta \mid \lambda_0, \underline{x}) \\[2mm]
&= \min_{(\lambda_0,\underline{x}) \in B_\alpha(\lambda_0^\star) \times B_\alpha(\underline{x}^\star)} \pi\left(\theta \mid s(\lambda_0, \underline{x})\right) \\[2mm]
&= \min_{\lambda_n \in B_\alpha(\lambda_n^\star)} \pi(\theta \mid \lambda_n) \ .
\end{aligned}
$$

In an analogous way the second equation is proved.

$\square$

Example 22.1: We continue example 21.1 from section 21.1 where exponential distributions are used as stochastic model and gamma distributions as conjugate family of *a priori* distributions. For precise data $\underline{x} = (x_1, \ldots, x_n)$ using the notation from example 21.1 we obtain

$$
\lambda_n = s(\lambda_0, \underline{x}) = \begin{pmatrix} \nu_0 + n \\ \beta_0 + \displaystyle\sum_{i=1}^n x_i \end{pmatrix} .
$$

In case of non-precise *a priori* hyperparameter $\beta_0^\star$ characterized by its α-cut representation

$$
\left(B_\alpha(\beta_0^\star); \ \alpha \in (0,1]\right),
$$

the α-cuts of the non-precise *a posteriori* hyperparameter $\beta_n^\star$ are given by

$$
B_\alpha(\beta_n^\star) = \left[\min_{\beta_0 \in B_\alpha(\beta_0^\star)} \min_{\underline{x} \in B_\alpha(\underline{x}^\star)} \left(\beta_0 + \sum_{i=1}^n x_i\right), \right.
$$

$$
\left. \max_{\beta_0 \in B_\alpha(\beta_0^\star)} \max_{\underline{x} \in B_\alpha(\underline{x}^\star)} \left(\beta_0 + \sum_{i=1}^n x_i\right) \right]
$$

$$= \left[\underline{B}_\alpha(\beta_0^\star) + \min_{\underline{x} \in B_\alpha(\underline{x}^\star)} \sum_{i=1}^{n} x_i, \quad \overline{B}_\alpha(\beta_0^\star) + \max_{\underline{x} \in B_\alpha(\underline{x}^\star)} \sum_{i=1}^{n} x_i \right].$$

If the minimum combination-rule is used, then the last expression simplifies to

$$B_\alpha(\beta_n^\star) = \left[\underline{B}_\alpha(\beta_0^\star) + \sum_{i=1}^{n} \underline{B}_\alpha(x_i^\star), \quad \overline{B}_\alpha(\beta_0^\star) + \sum_{i=1}^{n} \overline{B}_\alpha(x_i^\star) \right],$$

where $\underline{B}_\alpha(\cdot)$ and $\overline{B}_\alpha(\cdot)$ denote the lower and upper ends of the one-dimensional α-cut of $\beta_0^\star$ and $x_i^\star$, respectively.

The non-precise *a posteriori* hyperparameter $\lambda_n^\star$ generates a non-precise gamma distribution.

In figure 22.1 the characterizing functions of a sample of non-precise observations are depicted. For these data the *a posteriori* density is a non-precise density whose α-level curves are drawn in figure 22.2.

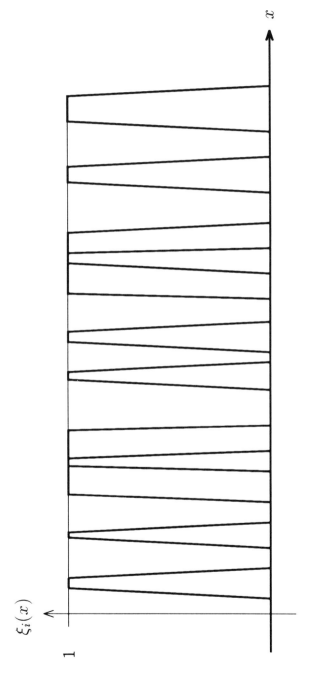

Figure 22.1: *Non-precise sample*

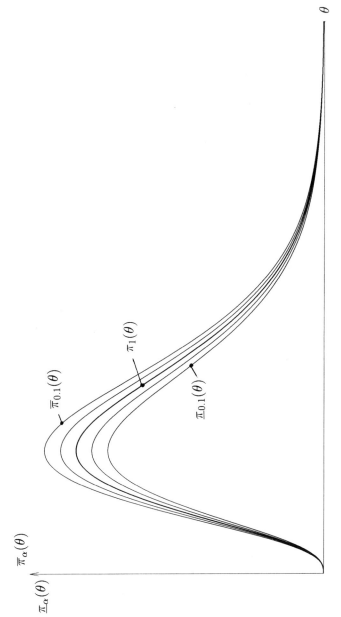

Figure 22.2: α-Level curves of the non-precise a posteriori density

Exercises

1. Explain the sequential information processing property of Bayesian updating in case of precise *a priori* density $\pi(\cdot)$ and precise data $x = (x_1, \ldots, x_n, x_{n+1}, \ldots, x_{n+k})$.

2. Explain the validity of

$$\underline{g}_{n,\alpha}(\theta) = \underline{g}_{n-1,\alpha}(\theta) \cdot \min_{x \in B_\alpha(x_n^*)} f(x \mid \theta)$$

and

$$\overline{g}_{n,\alpha}(\theta) = \overline{g}_{n-1,\alpha}(\theta) \cdot \max_{x \in B_\alpha(x_n^*)} f(x \mid \theta)$$

in case of using the minimum combination-rule.

3. Show in detail that for the exponential distribution the family of gamma distributions forms a conjugate family of *a priori* distributions.

4. Explain in detail the formula

$$\lambda_n = s(\lambda, \underline{x}) = \begin{pmatrix} \nu_0 + n \\ \beta_0 + \sum_{i=1}^{n} x_i \end{pmatrix}$$

in the example from this section (Continuation of example 21.1).

23 Bayesian decisions based on non-precise information

In Bayesian decision analysis based on precise *a priori* information and precise data decisions are based on the precise *a posteriori*
distribution of the parameter by minimizing the expected loss.

Let θ be the parameter, $\mathcal{D}$ the set of possible decisions d, and
$L(\cdot, \cdot)$ the loss function, i.e., $L(\theta, d)$ is the loss which occurs if the
parameter has value θ and the decision d is taken. The expected
loss based on the distribution $\pi(\cdot)$ of the parameter is

$$\mathbb{E}_{\pi(\cdot)} L(\tilde{\theta}, d),$$

where $\tilde{\theta}$ denotes the stochastic quantity which models the uncertainty concerning the parameter.

A *Bayesian decision $d_\circ$* in the class $\mathcal{D}$ of possible decisions is
given by

$$\mathbb{E}_{\pi(\cdot)} L(\tilde{\theta}, d_\circ) = \min_{d \in \mathcal{D}} \mathbb{E}_{\pi(\cdot)} L(\tilde{\theta}, d).$$

For non-precise distributions $\pi^\star(\cdot)$ — for example non-precise
a posteriori distributions $\pi^\star(\cdot \mid D^\star)$ — a generalization of the
concept of expectation is necessary.

For non-precise *a priori* distribution $\pi^\star(\cdot)$ of the parameter $\tilde{\theta}$,
the generalized *fuzzy expected loss*

$$\mathbb{E}^\star L(\tilde{\theta}, d)$$

is given via its family

$$\left(B_\alpha\left(\mathbb{E}^*L(\tilde{\theta},d)\right); \ \alpha \in (0,1]\right)$$

of α-cuts. These α-cuts are obtained from the fuzzy probability distribution $\pi^*(\cdot)$.

Denoting the α-cuts $B_\alpha\left(\mathbb{E}^*L(\tilde{\theta},d)\right)$ by

$$\left[\underline{\mathbb{E}}_\alpha^* L(\tilde{\theta},d), \ \overline{\mathbb{E}}_\alpha^* L(\tilde{\theta},d)\right],$$

the determination of the values of the endpoints of these intervals is described for discrete parameters and continuous parameters separately.

Remark 23.1: Alternatively, instead of loss functions, *utility functions* $U(\cdot,\cdot)$ can be considered. In this case Bayesian decisions are defined as the ones which maximise the expected utility

$$\mathbb{E}_{\pi(\cdot)}U(\tilde{\theta},d),$$

23.1 Discrete parameter space

If the parameter space consists of a finite number of different values, i.e., $\Theta = \{\theta_1,\cdots,\theta_k\}$, then the non-precise information

$$\pi^*(\cdot) \ \hat{=} \ \left(p_j^*, \ j = 1(1)k\right)$$

concerning the parameter can be given in form of k non-precise probabilities p_j^*, $j = 1(1)n$, where

$$B_\alpha(p_j^*) = \left[\underline{p}_{j,\alpha}, \overline{p}_{j,\alpha}\right]$$

denote the α-cuts.

For loss function $L(\theta,d)$ the generalized *fuzzy expected loss*

$$\mathbb{E}^*L(\tilde{\theta},d) = \sum_{i=1}^{k} L(\theta_i,d)p_i^*$$

is defined using the α-cuts $B_\alpha(p_j^*)$ by

$$\overline{\mathbb{E}}_\alpha^* L(\tilde{\theta}, d) = \sum_{j=1}^k L(\theta_j, d) \cdot \overline{p}_{j,\alpha}$$

and

$$\underline{\mathbb{E}}_\alpha^* L(\tilde{\theta}, d) = \sum_{j=1}^k L(\theta_j, d) \cdot \underline{p}_{j,\alpha}.$$

The characterizing function $\phi(\cdot)$ of $\mathbb{E}^* L(\tilde{\theta}, d)$ is given using proposition 2.1 by

$$\phi(x) = \max_{\alpha \in (0,1]} \alpha \cdot I_{B_\alpha(\mathbb{E}^* L(\tilde{\theta}, d))}(x) \qquad \forall \quad x \in \mathbb{R}.$$

23.2   Continuous parameter space

For continuously distributed parameter $\tilde{\theta}$ with fuzzy density $\pi^*(\cdot)$ and α-level curves

$$\underline{\pi}_\alpha(\cdot) \qquad \text{and} \qquad \overline{\pi}_\alpha(\cdot)$$

the interval endpoints of the α-cuts $B_\alpha\left(\mathbb{E}^* L(\tilde{\theta}, d)\right)$ of the fuzzy expected loss are

$$\overline{\mathbb{E}}_\alpha L(\tilde{\theta}, d) = \int_\Theta L(\theta, d) \, \overline{\pi}_\alpha(\theta) d\theta$$

and

$$\underline{\mathbb{E}}_\alpha L(\tilde{\theta}, d) = \int_\Theta L(\theta, d) \, \underline{\pi}_\alpha(\theta) d\theta.$$

The characterizing function of $\mathbb{E}^* L(\tilde{\theta}, d)$ is obtained in the same way as in section 23.1.

23.3　Models with hyperparameters

In case of a hyperparameter λ for the distribution $\pi(\cdot) = \pi(\cdot \mid \lambda)$ of $\tilde{\theta}$ we have, for discrete or continuous parameter

$$\mathbb{E}_{\pi(\cdot \mid \lambda)} L(\tilde{\theta}, d) = \sum_{j=1}^{k} L(\theta_j, d)\, p(\theta_j \mid \lambda)$$

and

$$\mathbb{E}_{\pi(\cdot \mid \lambda)} L(\tilde{\theta}, d) = \int_{\Theta} L(\theta, d)\, \pi(\theta \mid \lambda)\, d\theta$$

respectively.

$\mathbb{E}_{\pi(\cdot \mid \lambda)} L(\tilde{\theta}, d)$ is a function of the hyperparameter λ, i.e.,

$$\mathbb{E}_{\pi(\cdot \mid \lambda)} L(\tilde{\theta}, d) = h(\lambda).$$

If the imprecision of $\pi^{\star}(\cdot)$ is generated by the imprecision of a hyperparameter $\lambda^{\star}$, then the propagation of the imprecision is best calculated using α-cuts.

Let

$$\big(B_{\alpha}(\lambda^{\star});\ \alpha \in (0, 1]\big)$$

be the family of α-cuts of $\lambda^{\star}$, then the *generalized non-precise* value $\mathbb{E}^{\star} L(\lambda^{\star}, d)$ of the *expected loss*

$$\mathbb{E}^{\star} L(\lambda^{\star}, d) = \mathbb{E}_{\pi(\cdot \mid \lambda^{\star})} L(\tilde{\theta}, d)$$

is given by its family of α-cuts

$$\big(B_{\alpha}\big(\mathbb{E}^{\star} L(\lambda^{\star}, d)\big);\ \alpha \in (0, 1]\big)$$

which are given — using the function $h(\lambda)$ — by

$$B_{\alpha}\big(\mathbb{E}^{\star} L(\lambda^{\star}, d)\big) = \Big[\min_{\lambda \in B_{\alpha}(\lambda^{\star})} h(\lambda),\ \max_{\lambda \in B_{\alpha}(\lambda^{\star})} h(\lambda) \Big]$$

$$\forall\ \alpha \in (0, 1].$$

Remark 23.2: Non-precise expected values $\mathbb{E}^{\star} L(\lambda^{\star}, d)$ cannot always as easily be compared as precise numbers. Therefore, finding Bayesian decisions can make problems. These problems are similar to those connected with statistical tests for non-precise data.

The development of suitable models for a generalization of Bayesian decisions to the situation of non-precise distributions and non-precise data was partially done in fuzzy set theory. But from the viewpoint of statistics these generalizations are not satisfactory. Therefore, these problems form an interesting field for further research in statistical inference.

Exercises

1. Let $\{\theta_1, \cdots, \theta_k\}$ be the set of possible values for the parameter θ. The possible decisions are to select one of the values $\theta_1, \cdots, \theta_k$. Further assume that the non-precise information $\pi^*(\cdot)$ concerning the parameter is given by k fuzzy numbers $p_1^*, \cdots, p_k^*$. The loss function $L(\cdot, \cdot)$ measures the loss for selecting θ_i if the parameter is θ_j, i.e., $L(\theta_j, \theta_i)$. How are the characterizing functions of the fuzzy expected loss for every possible decision calculated?

2. For exponential distribution of X and conjugate family of *a priori* distributions as in example 21.1, calculate the non-precise expected values $\mathbb{E}^* L(\lambda^*, d)$ for given loss function $L(\theta, d)$.

Outlook

Research on statistical methods in face of non-precise data and general non-precise information is in progress and some important questions are still not solved. There are many interesting problems, for example statistical tests in case of non-precise data and general Bayesian decision rules in these situations.

Also, principal questions concerning the mathematical description of non-precise data and the determination of characterizing functions of these data in different applications are important research topics.

Details on yet unpublished research results in this field can be obtained from the author at the following address:

Institute of Statistics and Probability Theory
Technische Universitaet Wien
A-1040 Wien
Austria

References

[1]   H. Bandemer (Ed.): *Modelling Uncertain Data*, Akademie
       Verlag, Berlin, 1993.

[2]   H. Bandemer, W. Näther: *Fuzzy Data Analysis*, Kluwer
       Academic Publishers, Dordrecht, 1992.

[3]   F.P.A. Coolen: On Bernoulli experiments with imprecise
       prior probabilities, *The Statistician* 43, No. 1, 155-167 (1994).

[4]   F.P.A. Coolen: Bounds for expected loss in Bayesian deci-
       sion theory with imprecise prior probabilities, *The Statisti-
       cian* 43, No. 3, 371-379 (1994).

[5]   D. Dubois, H. Prade: Fuzzy sets and statistical data, *Euro-
       pean Journal of Operational Research* 25, 345-356 (1986).

[6]   S. Frühwirth-Schnatter: On statistical inference for fuzzy
       data with applications to descriptive statistics, *Fuzzy Sets
       and Systems* 50, 143-165 (1992).

[7]   S. Frühwirth-Schnatter: On fuzzy Bayesian inference, *Fuzzy
       Sets and Systems* 60 (1993).

[8]   M.A. Gil, N. Corral, P. Gil: The minimum inaccuracy esti-
       mates in χ^2 tests for goodness of fit with fuzzy observati-
       ons, *Journal of Statistical Planning and Inference* 19, 95-115
       (1988).

[9]   C. Goutis: Ranges of posterior measures for some classes
       of priors with specified moments, *Intern. Stat. Review* 62,
       245-256 (1994).

[10]   J. Kacprzyk, M. Fedrizzi (Eds.): *Combining Fuzzy Impreci-sion with Probabilistic Uncertainty in Decision Making*, Lec-ture Notes in Economics and Mathematical Systems, Vol. 310, Springer-Verlag, Berlin, 1988.

[11]   R.E. Kalman: Identification from real data, in: M. Haze-winkel, A.H.G. Rinnoy Kan (Eds.): *Current Developments in the Interface: Economics, Econometrics, Mathematics*, D. Reidel, Dordrecht, 1982.

[12]   A. Kandel: Fuzzy statistics and policy analysis, in: P.P. Wang, S.K. Chang (Eds.): *Fuzzy Sets, Theory and Applica-tions to Policy Analysis and Information Systems*, Plenum Press, New York, 1980.

[13]   G.J. Klir, T.A. Folger: *Fuzzy Sets, Uncertainty, and Infor-mation*, Prentice Hall, Englewood Cliffs, N.J., 1988.

[14]   R. Kruse, K.D. Meyer: *Statistics with Vague Data*, D. Reidel Publ., Dordrecht, 1987.

[15]   K.G. Manton, M.A. Woodbury, H.D. Tolley: *Statistical Ap-plications Using Fuzzy Sets*, J. Wiley & Sons, New York, 1994.

[16]   S.P. Niculescu, R. Viertl: A comparison between two fuzzy estimators for the mean, *Fuzzy Sets and Systems* 48, 341-350 (1992).

[17]   S.P. Niculescu, R. Viertl: A fuzzy extension of Bernoulli's law of large numbers, *Fuzzy Sets and Systems* 50, 167-173 (1992).

[18]   H. Rommelfanger: *Fuzzy Decision Support Systeme*, Springer-Verlag, Berlin, 1994.

[19]   S. Schnatter: On the propagation of fuzziness of data, *En-vironmetrics* 2, No 2, 241-252 (1991).

[20]   H. Tanaka, T. Okuda, K. Asai: Fuzzy information and de-cision in statistical models, in: M.M. Gupta, R.K. Ragade, R.R. Yager (Eds.): *Advances in Fuzzy Set Theory and App-lications*, North-Holland Publ., Amsterdam, 1979.

[21] R. Viertl: Is it necessary to develop a fuzzy Bayesian infe-
      rence, in: R. Viertl (ed.): *Probability and Bayesian Stati-
      stics*, Plenum Press, New York, 1987.

[22] R. Viertl: Statistische Analyse unscharfer Messungen in Zu-
      verlässigkeitsanalysen, in: *Zuverlässigkeit von Komponenten
      technischer Systeme*, VDI-Verlag, Düsseldorf, 1989.

[23] R. Viertl: Estimation of the reliability function using fuzzy
      life time data, in: P.K. Bose, S.P. Mukherjee, K.G. Ra-
      mamurthy (Eds.): *Qualitiy for Progress and Development*,
      Wiley Eastern, New Delhi, 1989.

[24] R. Viertl: Modelling of fuzzy measurements in reliability
      estimation, in: V. Colombari (Ed.): *Reliability Data Collec-
      tion and Use in Risk and Availability Assessment*, Springer-
      Verlag, Berlin, 1989.

[25] R. Viertl: *Einführung in die Stochastik mit Elementen der
      Bayes-Statistik und Ansätzen für die Analyse unscharfer Da-
      ten*, Springer-Verlag, Wien, 1990.

[26] R. Viertl: Statistical inference for fuzzy data in environme-
      trics, *Environmetrics* 1, 37-42 (1990).

[27] R. Viertl: On descriptive statistics for non-precise data, in:
      *Bulletin of the 48th Session of the International Statistical
      Institute, Contributed Papers, Book 2*, Cairo, 1991.

[28] R. Viertl: Zur statistischen Analyse von unscharfen Daten,
      in: *Beiträge zur Umweltstatistik*, Schriftenreihe der Techni-
      schen Universität Wien, Band 29, 1992.

[29] R. Viertl: On statistical inference based on non-precise data,
      in: H. Bandemer (Ed.): *Modelling Uncertain Data*, Akademie-
      Verlag, Berlin, 1993.

[30] R. Viertl, H. Hule: On Bayes' theorem for fuzzy data, *Sta-
      tistical Papers* 32, 115-122 (1991).

[31] H.J. Zimmermann, L.A. Zadeh, B.R. Gaines (Eds.): *Fuzzy
      Sets and Decision Analysis*, TIMS Studies in the Manage-
      ment Sciences, Vol. 20, North Holland, Amsterdam, 1984.

List of Symbols

$B_\alpha(x^\star)$ α-cut of the non-precise observation $x^\star$

d decision

d_0 Bayesian decision

$D^\star$ non-precise data

Ex_θ exponential distribution with parameter θ

$\mathbb{E}_{\pi(\cdot)}L(\tilde{\theta},d)$ expectation of $L(\tilde{\theta},d)$ based on the distribution $\pi(\cdot)$ of the parameter $\tilde{\theta}$

$\mathbb{E}^*L(\tilde{\theta},d)$ fuzzy expectation of $L(\tilde{\theta},d)$ based on the fuzzy distribution $\pi^*(\cdot)$ of the parameter $\tilde{\theta}$

$\exp(\cdot)$ exponential function

$f(\cdot \mid \theta)$ stochastic model, density

$\mathcal{F}(\mathbb{R})$ system of all non-precise numbers

$\mathcal{F}(\mathbb{R}^n)$ system of all non-precise n-dimensional vectors

$\hat{F}_n(\cdot)$ empirical distribution function

$\hat{F}_n(\cdot \mid x_1^\star,\ldots,x_n^\star)$ non-precise empirical distribution function

$F_n^\star(\cdot \mid x_1^\star,\ldots,x_n^\star)$ interval valued empirical distribution function

$g(\cdot \mid \underline{x})$ predictive density

$g^\star(\cdot)$ non-precise function

$\overline{g}_\alpha(\cdot)$ upper α-level curve of the non-precise function $g^\star(\cdot)$

$\underline{g}_\alpha(\cdot)$	lower α-level curve of the non-precise function $g^\star(\cdot)$
$g_n(\cdot \mid \lambda, \underline{x})$	non-normalized *a posteriori* density
$g^\star(\cdot \mid D^\star)$	non-precise predictive density
$\overline{g}_{n,\alpha}(\cdot)$	upper α-level curve of the non-normalized *a posteriori* density for non-precise data
$\underline{g}_{n,\alpha}(\cdot)$	lower α-level curve of the non-normalized *a posteriori* density for non-precise data
$\Gamma(\cdot)$	gamma function
$h_n(A)$	relative frequency of the event A
$h_n^\star(A)$	non-precise relative frequency of the event A
$\overline{h}_n(A)$	upper limit of the non-precise relative frequency
$\underline{h}_n(A)$	lower limit of the non-precise relative frequency
HPD	highest *a posteriori* density
$I_A(\cdot)$	indicator function of the set A
$K_n(\cdot, \ldots, \cdot)$	combination rule
$\kappa(X_1, \ldots, X_n)$	confidence function
ln	logarithmus naturalis (basis e)
$l(\theta; x_1, \ldots, x_n)$	likelihood function
Λ	space of tranformed parameters
M_X	observation space of the stochastic quantity X
M_X^n	sample space of the stochastic quantity X
$N(\mu, \sigma^2)$	normal distribution
$\mathcal{P}(M)$	power set of M
$\varphi(\cdot)$	characterizing function
$\pi(\cdot)$	*a priori* density

$\pi(\cdot \mid \lambda)$	*a priori* density with hyperparameter λ
$\pi(\cdot \mid \underline{x})$	*a posteriori* density for data $\underline{x}$
$\pi(\cdot \mid D)$	*a posteriori* density for data D
$\pi^\star(\cdot)$	non-precise *a priori* density
$\pi^\star(\cdot \mid D^\star)$	non-precise *a posteriori* density
$\pi^\star(\cdot \mid \underline{x}^\star)$	non-precise *a posteriori* density
$\overline{\pi}^\star_\alpha(\cdot \mid \underline{x}^\star)$	upper α-level curve of the non-precise *a posteriori* density
$\underline{\pi}^\star_\alpha(\cdot \mid \underline{x}^\star)$	lower α-level curve of the non-precise *a posteriori* density
$r_{x,y}$	empirical correlation coefficient
$r^\star$	non-precise empirical correlation coefficient
$S_n(\cdot)$	cumulative sum for non-precise samples
s_n^2	sample variance
$supp\big(\xi(\cdot)\big)$	support of the characterizing function $\xi(\cdot)$
$t(X_1, \ldots, X_n)$	statistic of the sample $X_1, \ldots, X_n$, test statistic
$t^\star$	non-precise value of a test statistic
$\tau(\theta)$	transformed parameter
θ	parameter
$\tilde{\theta}$	stochastic quantity describing the parameter
Θ	parameter space
$\Theta_{\underline{x}}$	confidence set
$\hat{\theta}^\star$	non-precise estimate for the parameter θ
$\vartheta(X_1, \ldots, X_n)$	estimating function for θ based on sample $X_1, \ldots, X_n$
$\underline{x} = (x_1, \ldots, x_n)$	vector of precise observations
$x^\star$	non-precise number, non-precise observation

$x_p^\star$ non-precise p-fractile

$\underline{x}^\star$ non-precise vector, non-precise combined sample element

$\xi_{x^\star}(\cdot)$ characterizing function of $x^\star$

Index

A

B

C